Not just the turkey thats roasting!

happy christmas Juan

love

Gillian

XXX 2007.

JM. She always gets it right!

40°s

Air & Space Disasters of the World

Dedication

To the casualties and their loved ones

Air & Space Disasters of the World

Xavier Waterkeyn

NEW HOLLAND

First published in Australia in 2007 by
New Holland Publishers (Australia) Pty Ltd
Sydney • Auckland • London • Cape Town

www.newholland.com.au

1/66 Gibbes Street Chatswood NSW 2067 Australia
218 Lake Road Northcote Auckland 0627 New Zealand
86 Edgware Road London W2 2EA United Kingdom
80 McKenzie Street Cape Town 8001 South Africa

National Library of Australia Cataloguing-in-Publication Data:
Waterkeyn, Xavier.
    Air and space disasters of the world.

    ISBN 978 1 74110 5506 (hbk.).

    1. Aircraft accidents. 2. Space vehicle accidents. I.
Title.

    363.12465

Publisher: Fiona Schultz
Production: Linda Bottari
Project Editor: Michael McGrath
Editor: Kirsten Chapman
Designer: Hayley Norman
Cover Design: Hayley Norman
Picture research: Xavier Waterkeyn, Michael McGrath
Printer: Imago

Photo captions for part openers

Part One, page 14—The fiery end of Germany's flagship Hindenberg was the curtain-call for airships and zeppelins.

Part Two, page 38—A heavily-armed hijacker at Beirut airport patrols in front a Royal Jordanian Alia Boeing 727, 11 June 1985, shortly after his group released all passengers and blew up the plane.

Part Three, page 86—An emergency evacuation slide (under the ladder) hangs from the wreckage of an Air France Airbus A340 that slid off the runway and caught fire at Toronto's Pearson International Airport in August 2005.

Part Four, page 166— A tractor (foreground) was towing an empty IranAir Airbus A-300 to a hangar at Mehrabad airport in Tehran when an Iranian Air Force Hercules C-130 crashed into the Airbus on take-off. Ten soldiers riding in the Hercules died in the accident.

Part Five, page 206—Iranian rescue workers inspect an airliner which caught fire on landing in Mashhad on 1 September 2006, killing 29 passengers. The Russian-made Tupolev suffered a tyre burst on landing and swerved off the runway.

Part Six, page 222—The remains of a wing from the Challenger space shuttle, dredged from the Gulf of Mexico.

Part Seven, page 234—Rescue workers spray a Philippine Airlines Boeing 737 with foam at Manila airport in May 1990. The brand new plane exploded as it was towed prior to takeoff, killing seven of its 110 occupants.

Part Eight, page 270— Firefighters, farmers and passengers try to put out the flames of a Garuda Boeing 737-400 after it crash-landed at Yogyakarta airport, 07 March 2007. While chemical foam is the firefighers' weapon of choice, when you're in a rice paddy you'll use whatever resources are available.

Part Nine, page 284—The reconstructed hull of TWA Flight 800. The Boeing 747 aircraft crashed 7 July 1996, killing all 230 aboard. The most expensive aviation investigation ever found no evidence that a bomb or missile brought the flight down.

Acknowledgements

With great thanks to Peter Gibson of the Australian Civil Aviation Safety Authority, to Deputy Director Information and Investigations Alan Stray, to Senior Transport Safety Investigators Robert Kells and Michael Watson of the Australian Transport Safety Bureau (Aviation Safety Investigation).

Thanks also to Kirsten Moonen and Denise Cupples of Borders Skygarden for helping out with some design issues and to Nick Bird who at least helped me a little bit, and to Mark Bird for paying all of Nick's bills.

A very special thanks to my editor Kirsten Chapman, to my project manager and picture researcher Michael McGrath, whose insight and eye for detail were invaluable, to my inspirational designer Hayley Norman and to Fiona Schultz for her dedication and staunch support.

PREFACE

Notes on terminology and aerodynamics

Flying and air navigation are in many ways the intellectual descendants of shipping and sea navigation. As such, aerial navigators have inherited several concepts and terms from their more water-bound colleagues. Here is a set of explanations of and simple conversions for the measurements used in this book, as well as a brief overview of how aeroplanes actually fly.

Time

The aviation industry uses 24-hour time. Midnight is 00:00 (pronounced zero hundred hours) and the following hours are 01:00 (zero one-hundred hours), 02:00 (zero two-hundred hours) and so forth. One pm is 13:00 (thirteen hundred hours), and the hours progress logically up to 23:59 (twenty-three fifty-nine). Where records show when something happened to the second, I have included the information in this format, for example 23:42:13 (thirteen seconds past 11:42 pm).

Please also note that black boxes use Coordinated Universal Time, which is based on Greenwich Mean Time and which, for reasons of historical compromise, has the abbreviation UTC. I have taken the liberty to convert the transcripts to local time to avoid confusion, unless stated otherwise.

Distance and speed

There are a number of different ways to measure distances.
1 kilometre = 0.62 statute miles = 0.54 nautical miles
Another way of looking at it is:
1 nautical mile = 1.15 statute miles = 1.85 kilometres
Airspeed is often given in knots (kts), which equal 1 nautical mile per hour. A term you may also see now and again is KIAS, which stands for Knots Indicated Airspeed and refers to the airspeed shown on the airspeed indicator.
1 knot = 1.152 miles per hour = 1.85 kilometres per hour

When dealing with critically short time spans it's often more significant to think in these terms:
1 knot = 1.69 feet per second = 0.51 metres per second.

You can appreciate then that if a 747-300 is cruising at a rate of 907 kilometres per hour (490 knots) it's covering 252 metres—just under four times its body length—every second or, to put it another way, a mile every seven seconds. If an accident happens, every second counts.

Please note that depending on the country in which the accident report was written, the original figures may have been in knots, miles per hour or kilometres per hour. For the sake of tidiness, I've rounded off the figures to the nearest ten except in those cases where I felt it was important to make finer distinctions.

Altitude

To make things more confusing, pilots and their colleagues often give the height of an aircraft in feet, usually thousands of feet. 1000 feet = 304.8 metres

So a plane cruising at 30 000 feet is about 9 kilometres or 5.68 miles up. Furthermore, professionals refer to plane altitudes by knocking off the last two zeros of the level in feet, so 'flight level 310' refers to a plane flying at 31 000 feet—well above the level at which humans can breathe comfortably.

If an object is in freefall it takes about 15 seconds to reach terminal velocity, which can be anything from 200 kilometres per hour to 320 kilometres per hour. So in a bad-case scenario, even if you survive being thrown out of an aeroplane that's falling apart in midair, you could hit the water or the ground at around 320 kilometres per hour (200 miles per hour).

Direction and turning

With reference to the direction in which an aeroplane is travelling, professionals refer to 'headings' and express turns in terms of the number of degrees clockwise from north. So a 'heading of 270' means 270 degrees clockwise from north, which is due west. A heading of due north is usually expressed as 360 rather than 0.

The front of a plane is called the nose, the back the tail. From the point of view inside the cockpit, the left wing is traditionally port, the right starboard; although people now often just speak of left and right. Engines are numbered from left to right, so that in a four-engined plane the outer left is engine one, the inner left engine two, the inner right engine three and the outer right engine four. In a three-engined plane the tail-mounted engine is engine two.

When the nose tilts up or down this is called 'pitch'. When the nose points left or right this is called 'yaw'. When the wings tilt up or down relative to the central axis along the plane's length this is called 'roll'. A 360-degree roll is a spin. The laws of aerodynamics mean that if you want to turn a plane left or right it needs to yaw and roll at the same time, this is called 'banking'.

Ageing aircraft fleets, vast distances and lax procedures conspire to give Africa the worst air safety record of all.

Flying and stalling

Aeroplanes work by using engine power to move through the sea of air that they fly in. However, even at sea level, the air is 800 times less dense than water, so it's a rather thin medium in which to travel. Nevertheless, if air is moving fast enough, or if you are moving fast enough through it, air seems to take on more substance, becoming something that you can really feel.

Planes are able to fly because of 'lift'. Wings have an upper curved surface and a lower straight surface. When air passes over the curved upper surface of a wing, it moves faster than the air under the wing, because the curved surface is a longer distance to go over than the straight underside. Not only does the air travelling over the top of the wing move faster, it is also less dense. This creates a vacuum over the upper surface of the wing. Nature likes to fill vacuums, if it can, and the nearest available vacuum filler is the body of the wing itself. So the wing is then sucked up into the vacuum that the wing itself creates, taking the rest of the plane with it—and, voila, you're flying.

A plane will continue to climb until its speed matches the forces creating this vacuum. It's all a balancing act: maintain this speed and the plane will stay at the same altitude; if the plane starts to slow down, the vacuum over the wing becomes less powerful and the plane starts to descend. However, if the plane drops below a certain critical airspeed, there won't be enough lift to keep the plane in the air—and, voila, you're stalling.

In effect, the plane will simply drop out of the sky and remind everyone that it is, in fact, heavier than the air around it.

I've listed the air crash narratives in broad chronological order, so that those who read it in this way can develop a feel for the technological and social changes that have taken place in the airline industry over the past hundred years. There are also special feature sections and boxes to highlight particular aspects of air disasters.

Currency

Unless stated otherwise, all amounts in this book are in US dollars, which is the international currency of aviation.

Space shuttle *Columbia* taking off on her final, ill-fated mission.

CONTENTS

206

222

234

284

INTRODUCTION

The cost of aeroplane crashes

IT'S DIFFICULT TO ESTIMATE how many people have died in aeroplanes or on the ground as a result of aeroplane activity. When we think about air disasters we tend to ignore combat situations; we accept that deaths from bombings or crashes during warfare are part of the much larger disaster that is war itself. Usually when we think plane crashes, we think civil aviation. We think about plane-loads of people who are just trying to travel from 'A' to 'B' with the minimum of fuss and the maximum of comfort. People like us who are simply going about their business and have no inkling that in a matter of minutes, their lives, hopes and dreams will be over—all because they caught the wrong flight at the wrong time.

Wrong is the key word here. The modern commercial jet is an extremely complex machine made up of millions of parts and costing tens or even hundreds of millions of dollars. Planes represent an enormous investment in science and technology, but flying is *inherently* risky. To many of us it just doesn't make sense that Boeing 747s with full payloads of fuel, cargo and passengers and weighing more than a quarter of a million tons each can even take off and stay in the air.

And, of course, it isn't a simple matter. The everyday miracle of commercial air transport depends on the proper working of millions of different machine parts, as well as the competence of the air crew, ground staff and control tower operators and the correct functioning of

their radar equipment. If any of these tightly coordinated elements fail, disaster can be only seconds away. That's why we have safety protocols, procedures, mechanisms, fail-safes, legislation and regulations that require the felling of whole forests to document. Otherwise things can become even more expensive than they are.

In the United States of America, in an average year, 100 aeroplane passengers lose their lives, compared to a world average of around 800 to 900 per year. This is really quite a good average, when you consider how much more frequently Americans fly than other nationalities. Another useful comparison is the rate of deaths per 10 million commercial passengers. The world regional rates for these are, on rounded averages:

North America: 7

Europe: 9

Australasia: 3

South America: 57

Africa: 130

Some years are worse than others, but a death in an air crash remains an extremely rare event, statistically.

In 1996 there were a total of 1187 major commercial aircraft fatalities. Compare this figure to some American statistics.

In 2005:

714 died in boating accidents

19 000 were murdered

41 907 died in car accidents

In an average year:

31 000 people (Americans) commit suicide

32 000 people die of an adverse reaction to a prescription drug

111 000 people die due to alcohol related causes

430 000 people die due to tobacco related causes

In Australia there have been no air fatalities in a major commercial airline accident involving a hull loss and the deaths of more than 100 people, ever. However, according to the Australian Transport Safety Bureau, there has been a total of 175 000 deaths since motorcars were introduced—equivalent to 15 000 deaths per year or one death every 35 minutes.

The launch of the De Havilland Comet ushered in the era of jet-propelled passenger aircraft—the Jet Age.

PART ONE
Air disasters before the Jet Age

THE FOLLOWING ACCIDENT HISTORIES ARE from the early decades of flight when air travel was a military imperative, a business necessity for those in a hurry or an expensive indulgence for the very rich. These were the days of propellers and airships, before the invention of the jet and the contemporary world of mass aerial transit.

17 September 1908

Flight: *Wright Flyer III*
Model: Flyer
Fatalities: 1/2

First Lieutenant Thomas Etholen Selfridge was the first person to die in an accident involving a heavier-than-air, powered, controllable aircraft. This historical event happened on the field of the newly created Aeronautical Division of the US Signal Corps at Fort Myer, Virginia, in the *Wright Flyer III*—piloted by no less a man than Orville Wright.

In the few years following the Wright brothers' historic breakthrough at Kitty Hawk, flying had become a phenomenon that spread like wildfire. Anyone who could was aching to take to the skies, now that Wilbur and Orville had shown the world that it was possible. However, flying would forever exact its price, so perhaps it is fitting that the dubious honour of being the first modern pilot responsible for a death on the wing should go to one of the fathers of aviation, and that the aeroplane involved should be a Wright plane—and not just any Wright plane, but a Flyer, the third of that name.

Thomas Selfridge graduated from West Point in the Class of 1903, the same year that Douglas MacArthur took first place. Selfridge's first flight in training was at Bras D'Or Lake, Nova Scotia, in the *Cygnet*, a kite designed by Alexander Graham Bell. This was the first flight in a heavier-than-air craft ever recorded in Canada, on 6 December 1906. A few months later Selfridge designed the *Red Wing*, the Aerial Experiment Association's first powered aircraft. The *Red Wing*'s maiden flight on 12 March 1908 was the first public demonstration of powered flight in America, but it crashed on its second flight only five days later. Selfridge was not the pilot; he was training to fly dirigibles (airships), specifically Thomas Baldwin's *Dirigible Number One*, which the US Army had purchased in July 1908.

As fate would have it, the army was interested in the military potential of aeroplanes and had agreed in principle to purchase a Wright plane. The collective decision was that Selfridge would be the passenger on Orville Wright's demonstration flight.

The flight seemed to go smoothly at first as the *Flyer III* circled Fort Myer four times, but at 17:18, halfway through circuit number five while the plane was at a height of 150 feet (46 metres) one of the propellers mounted on the rear of the *Flyer III* detached. As it spun out of its axle, it tore through the cables bracing the rudder. Although Wright managed to glide the plane down to about 75 feet (23 metres) it had lost too much airspeed.

In a later letter to his brother Wilbur, Orville wrote:

It was on the very first slow turn that the trouble began ... A hurried glance behind revealed nothing wrong, but I decided to shut off the power and descend as soon as the machine could be faced in a direction where a landing could be made. This decision was hardly reached, in fact, I suppose it was not over two or three seconds from the time the first taps were heard, until two big thumps, which gave the machine a terrible shaking, showed that something had broken ... The machine suddenly turned to the right and I immediately shut off the power. Quick as a flash, the machine turned down in front and started straight for the ground. Our course for 50 feet [15 metres] was within a very few degrees of the perpendicular. Lt. Selfridge up to this time had not uttered a word, though he took a hasty glance behind when the propeller broke and turned once or twice to look into my face, evidently to see what I thought of the situation. But when the machine turned head first for the ground, he exclaimed 'Oh! Oh!' in an almost inaudible voice.

A Wright Brothers bi-plane crash at Fort Myer, Virginia, which injured Orville Wright and killed his passenger, Lieutenant Thomas Selfridge—the first man to die in a powered aeroplane.

The plane had stalled and took a nosedive. It crashed within seconds. Orville Wright broke ribs, his pelvis and a leg. He would spend three months in hospital. Thomas Selfridge's skull was cracked, fractured against the wooden uprights of the plane's frame. The man who had seemed destined to be a great airman underwent neurosurgery, but he never regained consciousness and died later that evening. The army would delay the purchase of any plane until the following year.

In 1917 the army leased the mudflats around Lake St. Clair and in 1920 bought 600 acres. They named the field Selfridge and during World War II it became one of the premier airfields in the United States. The airfield has hosted countless air shows and thousands of pilots have trained there. It is still in use today.

Thomas Selfridge is buried at Arlington National Cemetery. He was 26 years old.

1878 & 14 December 1920

Flights: The aerial postmen and Aircraft Transport and Travel (AT&T)
Models: hot air balloon and Handley Page
Fatalities: 1/1 and 4/8 (2 passengers and 2 crew)

Here's a reasonable definition of commercial passenger aviation: a publicly or privately owned airline that offers scheduled or chartered services to passengers. At the dawn of commercial aviation history, however, operators weren't concerned with passengers. Flying wasn't particularly comfortable and the earliest regular services were mail runs. People were willing to pay a premium for fast delivery and mail pilots became the

most glamorous postmen ever. One in particular, Charles Lindbergh, later enjoyed the celebrity of a rock star, but Lindbergh might easily not have made it. All of the earliest commercial aeroplane fatalities involved lone pilots on mail runs.

One story precedes even the Wright brothers. The postmaster of Lafayette, Indiana, noticed that his city was not only almost due west of New York, but that it was also in the line of prevailing westerly winds. He decided to try an experiment. On 17 August 1859 he entrusted 51-year-old ballooning enthusiast John Wise with a mission to carry 123 letters and some circulars to New York. Eager to take on the challenge, Wise readied his balloon, the *Jupiter*, but found that the weather wasn't cooperating. In 32 degrees Celsius heat, Wise had to take his balloon up to 14 000 feet (4270 metres) before he found any wind at all, then it took him south. In five hours aloft he only covered about 50 kilometres (30 miles) and our intrepid aviator had to land near the town of Crawfordville, Indiana—a feat that *The Lafayette Daily Courier* sarcastically called 'trans-county-nental'.

In 1964 the Smithsonian Institution bought the one piece of mail to have survived from this, the world's first 'successful' airmail delivery. It was from a Mary A Wells to a WH Munn and reads: 'Dear Sir, Thinking you would be pleased to hear of my improved health I embrace the opportunity of sending you a line in this new and novel way of sending letters in a balloon.' Mary paid her premium: the 3 cents postage is worth the contemporary equivalent of about $72.

Wise survived this crash and later flew observation balloons in the American Civil War. He died in 1878, at the age of 71, in a balloon crash into Lake Michigan, making his the first death of an aerial postman.

The ruins of a German Zeppelin being inspected by French troops. Early dirigibles were unsafe, as was all early air travel, but World War I didn't make them any safer.

Motor car maker and air pioneer Charles Stewart Rolls—the Rolls of 'Rolls Royce'—became the first British pilot to lose his life.

He was flying a Wright biplane, which broke up in the air and crashed during Bournemouth Aviation Week, 12 July 1910.

Such historical cheating aside, the earliest pioneers of regular air services were the Europeans—specifically the Germans. The German Airship Transport Corporation (Deutsche Luftschiffahrts Aktiengesellschaft, or DELAG) was the world's first airline. Using aircraft built by the Zeppelin Corporation, DELAG was founded in 1909. On the 22 June 1910, LZ7 airship *Deutschland* began the first scheduled passenger run between Frankfurt, Baden-Baden and Dusseldorf. Unfortunately, on 28 June the *Deutschland* crashed in the Teutoburg Forest. DELAG was undeterred and went on to build the LZ127 *Graf Zeppelin*, the most successful airship ever built.

On 1 January 1914 the St. Petersburg-Tampa Airboat Line became the first company in the world offering a scheduled heavier-than-air passenger service, as pilot Tony Jannus ferried passengers, one or two at a time across Tampa Bay, Florida, in his Benoist XIV biplane seaplane. Tampa's ex-mayor Abram C Pheil won the auction for the first flight and paid $400 for the privilege ($35 000 in today's terms). Tony Jannus died shortly after this in 1916, while testing a Curtis H-7 for the Russians in the Black Sea. Rescuers never recovered the body of the 27-year-old aviation pioneer and entrepreneur.

However, the distinction of the first fatal air crash involving paying passengers belongs to the British company Aircraft Transport and Travel (AT&T). The company was founded by George Thomas and Geoffrey de Havilland, who initially designed and built warplanes, but decided in 1916 to take advantage of what they hoped would be the wave of the future: commercial passenger air travel. They were right, but they were also premature.

By the end of World War I, AT&T was making progress. The year 1919 saw the establishment of the British Department of Civil Aviation and the opening of Hounslow Heath, London's first airport. More famous then as the historic haunt of highwaymen, Hounslow Heath aerodrome is almost forgotten now, but the privately owned Great Western Aerodrome to the north eventually evolved into Heathrow, the world's third busiest airport and Europe's largest. In 1919 AT&T also made its first flight on the cross-channel route Le Bourget to Hounslow and by the end of the year established a regular mail run between Hounslow and Paris. Two other airlines, Handley Page and Instone, soon started to operate alongside AT&T, also using converted bombers, mostly de Havillands, which was cheaper than building new, dedicated planes. Whatever early successes these companies had, though, they were short-lived.

On 20 December 1920 shortly after take-off in foggy weather, an AT&T-run Handley Page O/400 crashed at Cricklewood. The crash killed both of the crew and two of the six passengers. It was the end for AT&T. They made their last flight the day after. By February 1921 every commercial aviation company in the UK had gone under. What saved the British air industry was government sponsorship and a merger.

In March 1921 what was left of AT&T combined with British Marine Air Navigation, Handley Page Transport and Instone Airlines to create Imperial Airways. In 1939 Imperial merged with British Airways Ltd to form the British Overseas Airways Corporation (BOAC). BOAC was the first airline in history to use jet aircraft and in 1974 BOAC and British European Airways merged to form British Airways.

After the 1920 crash Geoffrey de Havilland went back to designing and manufacturing planes rather than trying to run an airline. He bought the parts of AT&T that didn't go into the merger and renamed the company de Havilland. De Havilland would be responsible for making some of the most famous aircraft of the early twentieth century and introduced the world to commercial jet travel in 1952. However, all of that lay in the future; the world had not yet lost faith in balloons.

24 August 1921 & 5 October 1930

Flights: The R38 and R101 airships
Models: R38 and R101 airships
Fatalities: 44/49 (R38) and 48/54 (R101)

In the early decades of the 20th century it seemed that mass passenger transport would belong to the airship. Planes were small, noisy, smelly, energy expensive and their size made them vulnerable to the weather. Airships, at their very fastest, would only ever reach speeds of 150 kilometres per hour (93 miles per hour), but early passenger aeroplanes weren't that much faster and couldn't accommodate anywhere near the number of passengers that a dirigible could handle.

Some did try. For example, the massive flying ships, the Dornier Do Xs, were the largest aeroplanes of their day. At 48 metres (157 feet) wingtip to wingtip, 40 metres (131 feet) long, and 10 metres (33 feet) high, they needed twelve engines to take them to a maximum—and rarely achieved—cruising speed of 175 kilometres per hour (109 miles per hour). In 1929 the first Do X carried 169 passengers on a one-hour flight, but, weighing in at well over 50 tons, it consumed fuel at a rate of 1500 litres (400 gallons) per hour. The technology of the 1920s just wasn't up to the challenge of mass air transit. Other planes of the era didn't even come close.

On the other hand, airships were large, quiet, clean, needed no expenditure of energy to stay airborne and they could carry scores of passengers at a level of comfort and luxury that aeroplanes can't even rival today. They could even fly at night; something that fixed-wing craft wouldn't be able to do safely for a long time.

Passenger airships really captured public imagination. Soaring and majestic, they were the ocean liners of the air. They simply oozed style. This was an age in which travel, at least recreational travel, was still the province of the very rich. Millionaire heiress and 'Pioneer Passenger' Clara Adams bought a ticket for a round trip on the *Graf Zeppelin* from Germany to America, just for the hell of it. As one of only 75 passengers, and the only woman, she paid $3000 for the privilege—at a time when $8 a week was a good wage, that equates to about $112 000 in today's terms. Clara no doubt thought it was money well spent in spite of the bad weather during the flight's 71 hours. No doubt the crowds, the confetti and the ticker-tape parade that greeted the airship's arrival made the passengers feel like celebrities. The White House even invited the crew to a reception.

In 1929 the LZ127 *Graf Zeppelin* topped itself again, completing a truly epic circumnavigation of the globe. It started on 8 August for the first leg from Lakehurst, New Jersey, to Friedrichshafen, Germany, then travelled across Siberia to Tokyo. The third leg took the 237-metre (776-foot) giant from Tokyo to San Francisco, then to Los Angeles, Chicago and back to Lakehurst on 29 August. The LZ127 travelled 31 400 kilometres (19 500 miles) in 21 days, 5 hours and 31 minutes.

FROM THE COCKPIT

At midnight the R101 sent the following transmission:
2400GMT. 15 miles SW of Abbeville speed 33 knots. Wind 243 degrees 35 miles per hour. Altimeter height 1500 feet. Air temperature 51 degrees Fahrenheit. Weather intermittent rain. Cloud nimbus at 500 feet. After an excellent supper our distinguished passengers smoked a final cigar and having sighted this French coast have now gone to bed to rest after the excitement of their leave-taking. All of the essential services are functioning satisfactorily. Crew have settled down to watch-keeping routine.

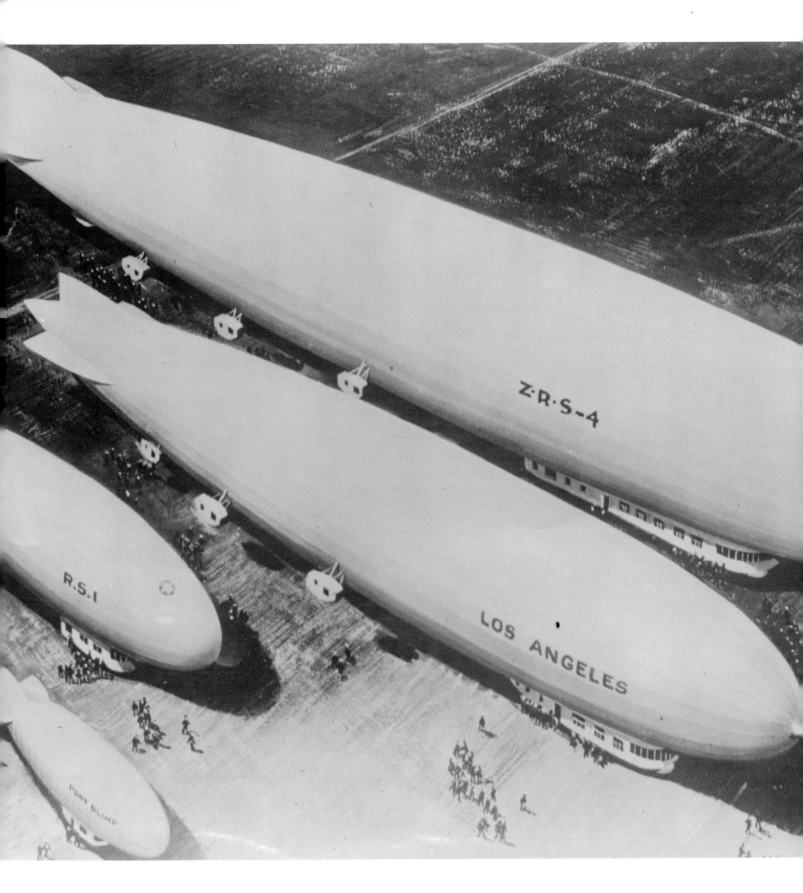

Airships came in one basic shape, but many sizes. The ZRS-4—better known as the USS *Akron*—ended its two-year career in the US Navy on 4 April 1933, when it crashed off the New Jersey coast, killing 73 of the 76 crew. The wreckage is still visible today, over 30 metres below the ocean surface east of Atlantic City.

A 1916 seaplane bi-plane, the first model to be produced by the Boeing aircraft company.

The Zeppelin Corporation's record of more than 2000 local flights and over 35 000 passengers carried between 1910 and 1914, without accident, was humiliating enough for the British, but the German achievements of the 1920's convinced Secretary of State for Air, Sir Samuel Hoare, that something had to be done.

The British had also had to make up for an embarrassing airship accident of 24 August 1921. On that day the British R38 had been on only its fourth flight, from Howden, Yorkshire, to Pulham Market, Norfolk. At 17:35 and travelling at 97 kilometres per hour (60 miles per hour) over the city of Hull, the crew attempted a sharp turn. The stress of the turn caused the R38 to break in two in midair. The nose and tail drooped before a fire engulfed the bow, and the largest airship in the world at the time crashed into the Humbar Estuary. Of the crew, 28 British and 16 Americans died.

In an effort to save face, albeit belatedly, the British ordered two new airships—the R100 and the R101. The Airship Guarantee Company built the R100 as a private venture. Among the people working on the R100 was the designer and engineer Barnes Wallis, of 'dam-buster' fame, and the stress engineer, Nevil Norway, who later became better known as the novelist Nevil Shute. Meanwhile the British Government set up the Royal Airship Works at Cardington to build the R101.

Construction of the R101 began in 1926, but problems and failed experiments delayed completion until 1929. The R101 frame eventually reached 777 feet (237 metres), just one foot longer that the *Graf Zeppelin*, but the ship was much larger at 156 000 cubic metres (5.5 million cubic feet), compared to the *Zeppelin*'s 105 000 cubic metres (3.7 million cubic feet). The frame was steel, but the gas bags that held the hydrogen to keep the ship aloft were made of goldbeater's skin —a membrane of bullocks' intestines.

While the team on the R101 were still working out bugs, the R100 launched in December 1929. On 29 July 1930 the R100 left for Saint-Hubert, Quebec. Seventy-eight hours later it arrived in Canada to massive public

acclaim. Some estimates are that as many as one million visitors came to see it. The R100's return trip to Cardington Airfield took only 58 hours and it arrived on 16 August.

The R101 was in danger of being grandstanded in its own country. Perhaps that's why the new State Minister for Air, Lord Christopher Thompson, pushed to start the project. The R101's maiden flight was supposed to be on 25 September, but bad weather delayed it until 4 October. At 18:34 it departed Cardington for Karachi, India. On board, under the command of Flight Lieutenant Carmichael Irwin, were some of the crème de la crème of the British aviation industry, including Sir Sefton Brancker, Director of Civil Aviation, and Lord Thompson himself.

By 20:21 the R101 was over London and sent this message: 'All well. Moderate rain. Base of low clouds 1500 feet. Wind 240 degrees [west south west] 25 miles per hour. Course now set for Paris. Intend to proceed via Paris, Tours, Toulouse and Narbonne.' The R101 had had to dump around 4 tons of water ballast to become airborne, and the crew was relying on rain to recapture ballast.

By 02:00 on 5 October the R101 was flying east to the town of Beauvais. According to the few survivors, at this point the airship's nose suddenly dipped. Rain had soaked the upper part of the nose, causing the outer skin to tear. The exposed skin bags leaked in the gusting winds. In 90 seconds the airship dipped 18 degrees. The crew compensated by pulling the elevators as far as they could go, but still only managed to raise the nose 3 degrees above horizontal. The ship was losing altitude. It was now down to 530 feet (162 metres). The crew had to release emergency ballast, but it was in the nose of the ship and they couldn't access it from the control room.

As the ship kept sinking, a crew member, Rigger Church, tried to reach the nose, but the ship crashed before he could reach it. The ship touched the ground while still moving at around 22 kilometres per hour (14 miles per hour) and travelled a further 20 metres (60 feet) before coming to rest. The starboard propeller

'Le Vaillant' passenger plane operated by Air France, circa 1937.

scratched the ground and the engine struts twisted from the main body. As the hot engine made contact with the body of the airship, the skin caught fire, and the hydrogen in the gas bags exploded successively from fore to aft.

Rigger Church escaped the burning ship, but died from his injuries three days later. Three engineers and another rigger were saved when the water ballast tanks above their engine cars burst, protecting them from the flames and allowing them to escape through the windows. Foreman Engineer J Harry Leech escaped through a break in the wall of the smoking room; tearing through the fabric of the ship, he ended up in the branches of a tree, wet with rain. Wireless Operator A Disley also tore his way free.

The victims of the crash, including Lord Thompson and Sir Sefton, were given a public service with full honours at Westminster and laid in state. All of the 48 were buried on the grounds of the Cardington Village churchyard. A memorial tomb dominates the small space today.

The remains of the R101 were left where they fell for months. Eventually scrap merchants salvaged what they could of the steel frame and the Zeppelin company bought 5000 kilograms (11 020 pounds) of duraluminium alloy structural components. It's possible that Zeppelin melted down the metal and reused it for the building of the *Hindenburg*.

In the light of the disaster, Thompson's successor, William Mackenzie, killed the British airship program. All of the R100's flights were cancelled and the government sold the R100 for scrap. A ship that had cost hundreds of thousands of pounds in research and development to build sold for £427 pounds (equivalent to £75 000 today or $140 000). History has judged that politicians drove the premature launch of the R101 before engineers had solved all its technical problems. As foreman engineer and survivor Harry Leech would later say: 'One thing that R101 proved was that politics and experimental work don't mix.'

9 February 1937

Flight: United Airlines Flight 23
Model: Douglas DC-3A
Registration: NC16703
Fatalities: 11/11 (9 passengers and 2 crew)

United Airlines flight 23 was making its final approach to San Francisco. At 20:50 the plane was about 3 kilometres (2 miles) offshore when the co-pilot accidentally dropped his microphone. It lodged between the elevator control column and the seat rail, jamming the controls. The crew was unable to remove the microphone as the plane commenced a rapid and disastrous dive into the bay.

A TRAGIC MILESTONE

United Airlines Flight 23 was the first accident involving a Douglas DC-3, a fixed-wing propeller-driven aircraft.

6 May 1937

Flight: The *Hindenburg*
Model: LZ129 airship
Fatalities: 35/97 (13 passengers, 22 crew) and 1 on the ground

Herb Morrison was probably expecting an easier job when his radio station WLS Chicago asked him to cover the *Hindenburg* landing, but his heartfelt account of the destruction of the *Hindenburg* before his eyes (see box page 28) and his pleading to bystanders to 'Get out of the way!' remain the most poignant tribute to the most dramatic destruction of an airship ever and to the 36 people who died that day in Lakehurst, New Jersey.

The pride of the German airship fleet was in every way the technical and physical superior of anything that the British had ever built, but in its own way, it too died of politics.

' ... it had staterooms for 72 passengers—cabins fitted with wardrobes, washbasins, luxurious carpeting and drapery, and real beds.'

It was the peak of contemporary airship construction, the brainlovechild of Doctor Hugo Eckener and Count Ferdinand von Zeppelin, the fathers of the German airship program.

Zeppelin had been interested in guidable balloons all his life. He'd made trips to America during the civil war and later in the 1870s to learn all that he could. He built his first airship in 1899 out of his own money—a machine that created so much interest at public demonstrations that he was able to finance a second ship from donations and the sale of lottery tickets. Although he died in 1917, his successor Hugo Eckener carried on his work. Eckener's crowning achievements were the *Graf Zeppelin*, the most successful airship ever built, and the *Graf Hindenburg*—the largest and the most luxurious.

Eckener detested the Nazi regime and hated how they used the *Hindenburg* for propaganda. Hitler eventually declared Eckener a 'non-person', but he was too valuable to dispose of. The LZ129 *Hindenburg* was simply too

awesome to ignore, and if Hitler wanted more of them, he needed Eckener.

At 245 metres (804 feet) in length, the LZ129 was almost as long as the *Titanic* and displaced almost 200 000 cubic metres (7 million cubic feet) of volume. It needed masterful control and a small army of crew to control her on the ground. You've only to look at newsreel footage of the day to see that airships were awe-inspiring, but a little hard to handle. Whatever its shortcomings the LZ129 mesmerised the world.

Built in 1935 for £500 000 pounds (£87 million or $165 million today) it had two main decks. The lower 'B' deck provided accommodation for 61 staff, as well as kitchens, baths, showers and a ballroom. The upper 'A' deck boasted a promenade and a lounge with a grand piano. By 1937 it had staterooms for 72 passengers— cabins fitted with wardrobes, washbasins, luxurious carpeting and drapery, and real beds. On its maiden flight to America, the *Hindenburg* carried 51 passengers

WORDS OF A HINDENBURG WITNESS

It's practically standing still now. They've dropped ropes out of the nose of the ship, and it has been taken ahold of down on the field by a number of men. It's starting to rain again; the rain had, uh, slacked up a little bit. The back motors of the ship are just holding it, uh, just enough to keep it from ...

It's burst into flames! It's burst into flames and it's falling, it's crashing. Watch it! Watch it now! Get out of the way, get out of the way! Get this, Charlie, get this, Charlie! It's crashing and it's crashing! It's crashing, terrible! Oh, my! Get out of the way, please! It's burning, bursting into flames and it's falling on the mooring mast, and all of the folks between. Oh, this is terrible. This is one of the worst catastrophes in the world! Oh Jesus, goodness gracious. Oh, four to five hundred feet into the sky. It's a terrific crash, ladies and gentlemen, the smoke and it's flames, now, and the frame is crashing to the ground, not quite to the mooring mast. Oh, the humanity, and all of the passengers screaming around here! I told you, I can't even talk to people whose friends are on there. It's, it's, it's, uh, I can't talk, ladies and gentlemen. Honestly it's a, a mass of smoking wreckage. And everybody can hardly breathe and talk ... I'm sorry, I can hardly breathe. I'm gonna step inside where I cannot see it. It's terrible ... Listen, folks, I'm gonna have to stop for a minute because this was the ... the worst thing I've ever witnessed.

Bystanders watch helplessly as the *Hindenburg* disintegrates in a ball of flaming hydrogen.

The wreckage of a De Havilland Dragon, which crashed in 1938. The Dragon was successful as a short-haul, low capacity airliner and was in service world-wide between the wars.

for 61 hours and 38 minutes. At $400 ($6000 in today's terms) for a one-way trip from Germany to America, passengers expected a little pampering.

In 1936, its first year of business, the *Hindenburg* flew 308 323 kilometres (191 583 miles) and ferried 2798 passengers and 160 tons of freight and mail. In one memorable appearance it glided over the opening ceremony of the 1936 Berlin Olympics—trailing the Olympic flag from its tail. In every sense the *Hindenburg* was a star. Hitler believed it would herald a new dawn of German domination of the skies.

On 3 May 1937 the *Hindenburg* left Germany for what was to be the first of eighteen scheduled voyages for the year. It made the Atlantic crossing in just 16 hours, spent the day of 5 May flying over New York and was expected to touch down at 08:00 on 6 May, but bad weather kept the *Hindenburg* in the air. It was hard enough to land in perfect weather; during a storm it was a nightmare. By 18:00 the rain had eased to a light drizzle. Friends and relatives, who had been waiting all day, were ready to welcome the passengers and a ground crew of 110 naval personnel and 138 civilians were waiting to bring the *Hindenburg* to tether.

At 19:19, as the *Hindenburg* neared her mooring mast, Captain Max Pruss ordered the water ballast to be dropped. The *Hindenburg* lowered 70 metres (200 feet). At 19:21 landing ropes came down and the steel mooring cable emerged from the nose. At this point the ship was 25 metres (75 feet) off the ground. Passengers were waving to the crowd below.

Then mechanic George Haupt saw a fire in gas cell number four in the ship's stern. From the outside, the crowd could see a flash of fire suddenly spring from the top rear of the dirigible. Captain Pruss made the decision to let the ship sink sternward to allow as many passengers as possible to escape. It took only 37 seconds for the ship to burn. Photographer Murray Becker of Associated Press described it as a 'moment of spectacular madness.'

In those few seconds the *Hindenburg* fell to earth, bounced up again 30 metres (100 feet), then fell again.

Many of those who were killed died because they jumped too early or mistimed their jumps. They fell to their deaths before the flames ever reached them. Others survived by sinking into the wet sand of the landing field while the fire raged above them. Others simply ran and could not account for how they escaped. As it was, it seemed a miracle that anyone survived at all. Captain Pruss survived, the entire back of his uniform burned away. He was one of the last to leave the ship and three US Navy men had to physically restrain him from going back into the burning wreckage to help more survivors. Later, back in Germany, he would argue and campaign to keep the German airship program going —to no avail.

The world lost faith in the airship after the *Hindenburg* went down. People still weren't used to air disasters and in spite of the high survival rate, the reality of physics—that hydrogen is highly combustible—could not compete with the newsreel footage of that massive fireball. The perception was, understandably, how could anyone survive that?

In Germany there was considerable anger against America. The *Hindenburg* had originally been designed to use the slightly heavier, but totally non-combustible gas helium for its lift. North America, however, had a virtual monopoly on helium and in a spirit of military caution had refused to sell it to the Nazi regime. Many felt that if America had let the *Hindenburg* use helium the disaster would never have happened. For them politics had taken precedence over the *Hindenburg's* safety.

Ironically, the worst ever airship disaster was in a helium-filled craft. The downing of the US Navy's USS *Akron* had claimed the lives of 73 men three years earlier, as well as being involved in a fatal accident prior to that. On 11 May 1932 while en route to her base at Sunnydale, California, the *Akron* made a stop at Camp Kenny, California. Camp Kenny lacked both the specialist mooring equipment and trained ground crew. The ship was light; she'd used 40 tons of fuel crossing the country and sunlight had caused the helium in her cells to expand, making her even more buoyant. The ground crew struggled, but she became impossible to control. In an effort to prevent her nosediving, the mooring cable was cut. Most of the ground crew had let go their lines, but as the *Akron* suddenly ascended, one man rose 15 feet (5 metres) before he let go and broke his arm. Three others went even further up. Both Aviation Carpenter's Mate Robert H Edsall and Apprentice Seaman Nigel M Henton couldn't hold on and fell to their deaths. Apprentice Seaman CM Bud Cowart managed to secure himself and hung on for one hour, before the crew of the *Akron* was able to hoist him onboard. Cameramen on the scene managed to capture the whole event on film.

Almost a year later, on the evening of 3 April 1933 the two-year-old *Akron* was flying down the coast of New England in severe and deteriorating weather. By 00:30 on 4 April immensely powerful storm winds flung her towards the sea and she crashed tail-first into the Atlantic. Fortunately, a German ship, the *Phoebus,* was in the vicinity and managed to rescue nine men, of whom only three survived. Tragically, a navy blimp, the J3 sent out in the search for the *Akron* also crashed, and two more died.

Although the *Akron's* crash claimed twice the lives of the *Hindenburg,* it wasn't quite the same. The *Akron* was a military vessel, it had been caught up in a huge storm, there was no fire and more importantly, there was no footage of the accident. It wasn't a peaceful palace in the sky, landing perfectly smoothly as it had at the end of twenty other safe trans-Atlantic journeys, before suddenly and inexplicably bursting into flames.

In the midst of the propaganda fiasco, the German Luftwaffe (air force) Chief Hermann Goering ordered the end of the German airships. One month after the destruction of the *Hindenburg,* he retired the LZ127 *Graf Zeppelin*—with its perfect safety record—and turned it into a museum. When the war came, not even the LZ127 was safe, and Goering ordered the destruction of all of the extant dirigibles to supply aluminium for the air force.

Theories abounded as to why the *Hindenburg* burned. Hugo Eckener thought that someone had sabotaged her, but he later stuck to theory that electrical brush discharge, also known as St Elmo's Fire, had built up in the storm conditions, sparked and set fire to gas cell number four, which had somehow sprung a leak. Other theories include fuel leaks and fabric fires, but the evidence is long gone. Curiously, although there were about fifteen newsreel cameramen and many other photographers at the site on the day, none of them captured the first instant of the *Hindenburg's* fire. In many cases they watched in shock, their cameras pointing to the ground, until they came to their professional senses and started shooting again.

At 16:30 on the day after the tragedy NBC broke its own rule about airing recorded news accounts. The recording became the first ever coast-to-coast radio broadcast. Herb Morrison went on to a long and respected career in radio. He died, aged 83, in 1989. After the *Hindenburg* disaster, Hugo Eckener kept a low profile. Miraculously he survived Hitler and died on 14 August 1954, four days after his 86th birthday.

The end of the rigid-framed airships was not, however, the end of the dirigible. As one final irony the US Navy continued to use them, but put more faith in the smaller, non-rigid blimps. At their peak during World War II, fourteen blimp squadrons found a role in naval reconnaissance. Together they covered the Atlantic, Pacific and Mediterranean and guarded 89,000 surface vessels. So effective were they at detecting German U-boats over the Straight of Gibraltar, that the Germans gave up the passage altogether. In today's world of lighter, stronger and safer materials, computer navigation, guidance technology and ever-increasing fuel costs, perhaps the airship will soon make a comeback.

The glamorous image of fixed-wing air travel for the well-heeled was in stark contrast to the noisy, bumpy, cold and uncomfortable reality. Air sickness could be so bad that a crash might seem—fleetingly—a preferable option.

A Swedish Fokker ski-plane, flown by Einar Lundborg had been sent to retrieve the survivors of Umberto Nobile's ill-fated *Italia* airship expedition to the North Pole. It crashed on landing, but came in useful as shelter. The *Italia* crashed on 25 May 1928, returning from a flight over the North Pole. Subsequent rescue attempts resulted in the death of the explorer Roald Amundsen.

'During Douglas trials of the innovation, however, two of their test pilots were almost overcome by CO_2 gas, but Douglas suppressed the information in their report to the CAA ...'

28 July 1945

Flight: military aircraft
Model: B-25D Bomber
Registration: 43-0577
Fatalities: 3/3 (1 passenger and 2 crew)
and 11 on the ground

Pilot Lieutenant Colonel William Franklin Jr was flying over Manhattan, in fog, attempting to reach Newark Airport, when air traffic control told him he'd have to wait three hours for clearance. Franklin then contacted La Guardia Airport, who wanted him to land because of the poor weather, stating that they 'couldn't see the top of the Empire State Building'. Franklin decided to descend for better visibility and found himself over the city, with skyscrapers all around him. He managed to narrowly avoid several buildings, but soon ran out of luck.

At 09:29 the 12-ton B-25, flying at 360 kilometres per hour (225 miles per hour), smashed into the 79th floor of the Empire State Building and straight into the offices of the War Relief Services and the Catholic Welfare Conference. The impact created a hole 6 metres (30 feet) wide. Fuel from the ruptured tanks started fires on two floors in which ten people died immediately, and one later in hospital.

One detached engine rammed through partitions and two firewalls before exiting the other side of the building and crashing into an adjacent building on 33rd Street. The other engine smashed into a lift shaft and cut the lift cables. The lift, containing an operator and a passenger, plummeted over 300 metres (1000 feet). Both women suffered severe injuries but survived.

While the bulk of the wreckage burned inside the Empire State Building, most fragments fell onto the building's fifth floor roof, set back from the pavement. A total of 26 people were injured as a result of the crash. The Empire State Building suffered $1 million worth of damage ($20 million in today's terms).

In similar circumstances, and also on a foggy Saturday in March 1946, Charles B Atlee was at work and on duty at the US Navy Officer's Discharge Centre on the 36th floor of the Manhattan Company Building on Wall Street. Both the road and the building were deserted, apart from a bank guard on the ground floor. At 21:00 a DC-3 bound for Newark crashed into the building. The fuselage flew into the building, as the engines and wings sheered off and tumbled to the ground. One fallen engine started a small fire in an adjoining building, the other engine buried itself in Wall Street. Fortunately, there was no explosion, but the crash killed all five onboard the aircraft.

24 October 1947

Flight: United Airlines Flight 608
Model: Douglas DC-6
Registration: NC37510
Fatalities: 53/53 (48 passengers and 5 crew)

While en route from Los Angeles to New York, United Airlines Flight 608 was crossing Bryce Canyon, Utah. The flight crew transferred fuel from one alternate tank to another, overfilling the second tank in the process. Due to a design flaw, the overspill leaked out of fuel venting line. Slipstream air currents then fed the ejected fuel back into the plane, through the fresh air intake of the cabin heater. When the crew turned on the heating, the fuel ignited, also setting off the emergency magnesium flares stored in the same compartment. At 12:29 the plane exploded and disintegrated, killing everyone onboard.

About three weeks later on 11 November, in an almost identical situation, American Airlines Flight 10 en route from San Francisco to Chicago reported a fire onboard and crash-landed in Gallup, New Mexico. This time, all of the 25 passengers and crew survived.

17 June 1948

Flight: United Airlines Flight 624
Model: Douglas DC-6
Registration: NC37506, *The Mainliner Utah*
Fatalities: 43/43 (39 passengers and 4 crew)

As a direct result of the fires on United Airlines Flight 608 and American Airlines Flight 10, the Civil Aviation Authority (CAA) ordered the installation of carbon dioxide fire extinguishers on all planes. During Douglas trials of the innovation, however, two of their test pilots were almost overcome by CO_2 gas, but Douglas suppressed the information in their report to the CAA, and the only result was a warning in the flight manual.

United Airlines Flight 624, bound to New York from Chicago, was flying over Pennsylvania when a fire started in the cargo hold. The crew used the CO_2 extinguishers to put out the fire, but as the plane's nose tilted down for an emergency descent, the CO_2 flowed into the cockpit and asphyxiated the crew. The out-of-control plane crashed through high voltage power lines, and exploded onto a hillside at Mount Carmel, killing everyone onboard.

Officials inspect the scene where 28 people died when a DC-3 'Dakota' crashed into a hillside and caught fire at Mill Hill, England on 18th October 1950. In the early days air crash investigations—with primitive methods and the absence of black box data recordings—were often little more than guesswork.

AERIAL TERRORISM IS OLDER THAN THE JET AGE and, according to the Aviation Safety Network Database, there have been over 1000 hijackings to date. The first recorded instance of air piracy happened in May 1930, when Peruvian revolutionaries skyjacked a Pan American mail plane. Their aim was to drop propaganda leaflets over Lima. The resourceful pilot, Byron D Richards, somehow convinced the gunmen that it would be a good idea if he delivered his mail first. After landing, he simply refused to cooperate any further.

Frustrated, the revolutionaries just gave up. Amazingly Byron Richards was flying a 707 over 30 years later when another set of hijackers took over his flight. Byron had retained his negotiation skills and he managed to talk the men out of flying to Cuba.

PART TWO:
Terrorism and other crimes

Hijacking is almost, but not always, politically motivated. Where war is usually the province of nations or states, smaller disaffected groups or individuals may have neither the temperament for conflict nor the resources for full-scale war. Nevertheless, they can be, in their own terms, extremely effective and make a little go a long way.

What follows are some of the most significant examples of aeroplane incidents that we can reasonably call acts of terrorism. In some cases, the military have shot down defenceless commercial airlines because of mistaken identity—or intentional malice, depending on your politics or point-of-view. There are also rare examples of lone individuals who have committed acts of violence upon aircraft. In all of these cases, a crime was committed, regardless of whether or not it was political or born of negligence. The victims are just as dead whatever the cause.

28 March 1933

Flight: Imperial Airways
Model: Armstrong Whitworth Argosy II
Registration: G-AACI
Fatalities: 15 (12 passengers and 3 crew)
Principle cause: sabotage (suspected)

In the days before black box recordings, when air crash investigation methods and technologies were still primitive, it was extremely difficult to determine the exact cause of an air disaster. What follows is, as best as anyone has ever been able to determine, an account of the first fatalities on a commercial airliner due to sabotage.

The British Imperial Airways Argosy II was flying at an altitude of 4000 feet (1220 metres) over Belgium, when a passenger started a fire in the cabin in an attempt to commit suicide. Witnesses reported fire coming out of the tail section. The crew attempted an emergency landing, but the fuselage split in two at around 14:30. The passenger succeeded in killing himself—as well as fourteen other people when the plane crashed.

10 October 1933

Flight: United Airlines
Model: Boeing 247-D
Registration: NC13304
Fatalities: 10/10 (7 passengers and 3 crew)
Principle cause: sabotage

On the 10 October 1933 a United Airlines flight was preparing to leave Newark, New Jersey, for Cleveland, Chicago and Oakland, California. Witnesses reported seeing a man carry brown package onto the plane in Newark, but no one thought anything more about it.

At around 21:00, while flying over Indiana at an altitude of only 1000 feet (300 metres), the plane exploded, killing everyone onboard.

At that time the US Bureau of Investigation (BOI)

handled inquiries into air crashes. The bureau's chief Melvin Purvis made a statement, which reveals that authorities were beginning to acquire some experience in the interpretation of crash evidence:

'Our investigation convinced me that the tragedy resulted from an explosion somewhere in the region of the baggage compartment in the rear of the plane. Everything in front of the compartment was blown forward, everything behind blown backward, and things at the side outward … The gasoline tanks, instead of being blown out, were crushed in, showing there was no explosion in them.'

The BOI ultimately determined that the cause of the explosion was a nitro-glycerine bomb. The BOI was never able to identify or charge anyone in connection to the first proven act of commercial air sabotage. In 1935 the BOI became the FBI.

9 September 1949

Flight: Canadian Pacific Airlines
Model: Douglas DC-3
Registration: CF-CUA
Fatalities: 23/23 (19 passengers and 4 crew)
Principle cause: sabotage

Watch and jewellery salesman Albert Guay was unhappy in his marriage to Rita Morel, and wanted to marry nineteen-year-old waitress, Marie-Ange Robitaille. Rita also had a $5000 life insurance policy, and Albert was eager to collect. He hit upon the cunning plan of killing Mrs Guay with a bomb planted on her plane. He may have been inspired by two ex-convicts who had planted

A TRAGIC MILESTONE

Onboard the flight was Chicago resident Alice Scribner—the first stewardess ever to die in a plane crash.

Before baggage X-raying was standard, or even technically possible, it was all too easy to stow a crude bomb inside a suitcase.

Fortunately the vast majority of airline passengers have never been malicious, deluded psychotics.

Of the many terrorist acts involving aircraft very few have ended in outright disaster. Most end peaceably—especially when the skyjackers realise that they've painted themselves into a corner.

a bomb on a Philippine Airlines flight in May of that year; the explosion over the sea between Daet and Manilla killed all of the ten passengers and three crew. However, Guay planned to be cleverer.

He approached his friend, clockmaker Généreux Ruest to make a timer. Ruest then persuaded his sister, Marguerite Pitre to buy some dynamite, ostensibly 'to clear a field of tree stumps'. Pitre had already been involved in the affair when she had arranged for Guay to meet his mistress Marie-Ange. What was one or two more favours? Perhaps a share in the insurance might have made her more willing to help.

With the bomb assembled and packaged, Pitre took it to the airport at Quebec City to post to Baie-Comeau. The parcel just happened to be going on the same flight that Mrs Guay was taking, and Albert Guay had just happened to increase his wife's insurance to $15 000 (worth about $200 000 in today's terms).

The DC-3 took off five minutes late. As the plane was flying over Sault-au-Cochon the bomb exploded in the forward luggage compartment, with the inevitable tragic results—everyone onboard was killed, including four children.

The timing of the explosion was a little inconvenient for Guay et al. Had the plane left on time, it would have exploded over the Saint Lawrence River. The evidence would have sunk and a determination of the cause of the crash would have been impossible, considering the forensic capabilities of the late 1940s. Three days after the crash, Pitre attempted, but failed, to kill herself.

As it was, it took the authorities just two weeks to arrest Guay. Police arrested Ruest and his sister over seven months later. All three were found guilty on 23 counts of murder. Albert Guay was hanged on 12 January 1951. His last words were, '*Au moins, je meurs celebre*' (at least I die a celebrity). Généreux Ruest was also hanged and was so ill from tuberculosis that he had to be taken to the gallows in a wheelchair. On 9 January 1953 Marguerite Pitre became the thirteenth woman to be hanged in Canada; she was also the last.

1 November 1955

Flight: United Airlines Flight 629
Model: Douglas DC-6B
Fatalities: 44/44 (39 passengers and 5 crew)
Principle cause: sabotage

John 'Jack' Graham's father died when Jack was only three. Unable to care for him, his mother Daisie put him in an orphanage and Jack spent the next ten years living in foster homes. Daisie took custody of her son again in 1945, and he went through several jobs before stealing some payroll cheques and forging the company president's signature, which netted him several thousand dollars. He bought a convertible and fled to Colorado. On 11 September 1951 a patrolman in Texas tried to arrest Jack for transporting alcohol in his car. Jack escaped and the ensuing high-speed car chase ended when Jack crashed into a police roadblock.

In the meantime Daisie had made several shrewd real estate investments and established a successful drive-in restaurant in Denver. She was able to afford restitution to the business that Jack had swindled, and the nineteen-year-old was put on probation. Two years later Jack was married with two children, living in his mother's house and working in her restaurant, paying off the debt he owed her. Early in 1955 Jack's truck mysteriously ended up on a railway line, where a train crashed into it. Jack used the insurance to pay off some of his debt to his mother. When the restaurant suffered a mysterious explosion a little later, Jack used the insurance money he collected again to end the debt.

By November that year, Daisie was planning to visit her daughter in Alaska. Jack's wife Gloria later testified that Jack gave his mother a Christmas present to take away with her. Jack denied that any such present existed, but he did in fact give her a box, which Daisie put in her suitcase. When the suitcase was loaded at Stapleton Airport, Denver, it ended up resting in the plane's rear luggage pit number four.

At 18:29 United Airlines Flight 629 was cruising at a height of 4000 feet (1220 metres) when the bomb went off. The plane disintegrated, killing everyone onboard almost instantly as a fireball consumed the plane. Farmers in the area reported a sight like a shooting star on that clear, full-moonlit night. Fragments of aeroplane, luggage and passengers fell across the fields of northern Colorado. Some bodies formed craters a third of a metre (1 foot) deep from the force of their impact. Some of the plane fragments would not be unearthed until the 1970s.

Rescue teams and employees of United Airlines spent the night and following day picking up bodies. One body ended up on a haystack; many were still strapped in seats in sections of fuselage, but so burned that they were unrecognisable. The passengers included an 18-month-old boy and a young couple celebrating their first wedding anniversary. A crowd numbering thousands gathered at the sight as authorities took the remains to a temporary morgue in the state armoury in the nearby town of Greeley.

The scattered remains of UA 629 immediately indicated a midair explosion, and investigators soon discovered the remains of a bomb made from TNT and a timing device. Working on the theory that a disgruntled ex-United Airlines employee might have planted the bomb, investigators began a background check on all of the passengers. Part of that check was into their insurance, including the coverage that could be purchased from coin-operated vending machines at the time.

One passenger immediately stood out. In Daisie King's luggage they found newspaper clippings of her son's forgery charges in 1951. They also quickly determined that, not only was she insured to the value of $37 500, but that Jack Graham was set to inherit

Daisie's considerable estate, including the Denver restaurant that had suffered the mysterious explosion earlier that year.

Initially Jack was very cooperative with the police and let them search his house. The search revealed wiring similar to that used in the bomb, as well as further hidden insurance policies to the value of $40 000—worthless though, because Daisie hadn't signed them. On 13 November, with the evidence piling up around him, Jack finally confessed that he had blown up UA 629 with a bomb made up of 25 sticks of dynamite, a 90-minute timer, a battery and two blaster caps. He described planting the bomb in his mother's luggage and hurrying her off to the airport. Today, with frequent traffic delays and the lengthy process of boarding flight, his bomb might have exploded on the ground, with perhaps less devastating consequences. As it was, his homemade annihilator
was all too effective. Jack was formally charged with 44 counts of murder the next day.

The apparent indifference and coldness of Jack's confession as it was reported in the press stunned the nation. He seemed to relish the attention and the case sparked something in the public imagination. Here was that rarest of things, a mass murderer, alive and talking. The court proceedings were the first in history to have full television coverage.

In spite of a later attempt to withdraw his confession, the trial proceeded to its inevitable conclusion, Jack's story was simply too accurate to have come from the police coercion he later tried to claim. After deliberations lasting only 72 minutes, the jury found Jack guilty of first-degree murder. The death penalty was automatic.

In jail, Jack's psychological state rapidly deteriorated. In September 1956, while waiting for the outcome of an appeal, he tied two socks around his neck and tried

As a symbol of 'American Capitalist Imperialism' Pan Am was the most unlucky terrorist target of them all.
(See feature pages 72–73.)

to hang himself from the bars of his cell. In spite of some support from anti-death penalty groups, the general public had no sympathy for Jack Graham. It would be difficult to feel sympathy for someone who said about his victims that, 'As far as feeling remorse for these people, I don't,' he said. 'I can't help it. Everybody pays their way and takes their chances. That's just the way it goes.'

At 19:45 on 12 January 1957, guards made Jack change into a pair of shorts—a necessary precaution to prevent clothes from storing the cyanide fumes. As they strapped Jack to the chair of the gas chamber and put a black hood over his head, the prison chaplain said, 'I hope God will forgive you for your sins. Take it like a man.' Jack's final words were. 'Okay. Thanks, Warden!' The prison doctor pronounced Jack Graham dead at 20:08.

A typical hostage stand off. The stark simplicity of images from the media show only the tip of an iceberg of tense, behind-the-scenes negotiation.

CUBA and AMERICA

Although the 1962 Cuban missile crisis was the most famous example of the complex dance macabre between America, Cuba and the USSR, the stormy relationship between Fidel Castro's regime and a succession of US governments continued for decades, with Cuba's national carrier, Cubana de Aviación (CU), becoming a prime target. There were so many hijackings and counter hijackings between Cuba and America during the 1960s and 1970s that it became the long-running staple of black jokes among cartoonists and stand-up comedians. Skyjacking incidents reached a peak in 1969, but, surprisingly, hijacking wasn't even a crime in Cuba until 1970 and American authorities didn't install metal detectors in airports until 1973.

A comprehensive list of the incidents makes for rather tedious and monotonous reading, but CU's record makes Pan American's safety record (see page 70) look tame by comparison. CU's number of total casualties may not be as high, or the events as spectacular or as well covered in the press as those of the American airline, but their frequency makes you wonder what was going on in the minds of CU's passengers at the time—they must have felt as if they were risking their lives on every flight.

Here are just some of the lowlights, outlining the most important or serious cases.

1 November 1958

Five Cuban revolutionaries hijacked CU Flight 495 en route from Miami to Havana. Their aim was to land at Sierra Cristal, in eastern Cuba, to deliver guns to Fidel Castro's younger brother Raul. The plane started running out of fuel and attempted an emergency landing at Preston Airport, Cuba, but crashed into the sea. Sixteen of the twenty passengers were killed, along with all four crew.

29 October 1960

No fewer than nine skyjackers, including the co-pilot of the flight, skyjacked a CU Douglas DC-3 at Havana on 29 October 1960. The plane made it to Key West and the incident claimed one life.

8 December 1960

Five weeks later five men skyjacked another CU flight, but were thwarted when the pilot deliberately crashed the plane, resulting in one fatality.

27 March 1966

Armed with a pistol, Angel Maria Betancourt Cueto, the flight engineer of a CU scheduled flight from Santiago de Cuba to Havana, attempted to divert the plane to America. The pilot, Fernando Alvarez Perez, refused to risk the lives of his 97 passengers more than he had to and landed in Havana anyway. Cueto then killed Perez and air-steward Edor Reyes and seriously injured Co-Pilot Evans Rosales. Cuban authorities later caught Cueto and executed him for murder.

The year of 1968

There were at least sixteen non-fatal skyjackings in this year. Disaffected socialists in America wanted to go to Cuba, and disgruntled capitalists in Cuba wanted to go to America. Why couldn't the two countries have set up a swap program and saved the airlines, their passengers and their crews a lot of grief? It would have been an admission that neither version of politics was the best of all possible worlds, and in the political climate of the time, such an admission would have been impossible.

Many of the skyjackers in this year and the following years had a history of mental illness. Some ended up in mental institutions after their convictions.

'Why couldn't the two countries have set up a swap program and saved the airlines, their passengers and their crews a lot of grief?'

The year 1969

This was a bumper year for the skyjackers.

- January: at least nine hijackings and attempted hijackings.
- February: at least four incidents.
- March: at least three incidents.
- May: at least two incidents.
- April: no incidents—it must have been a skyjacker holiday month.
- June: at least four incidents.
- July: at least two incidents.
- August: at least three incidents.
- September: at least three incidents.
- October: at least two incidents.
- November: one incident.
- December: one incident.

Total for the year: at least 34 instances of successful or attempted skyjackings. Fortunately none involved fatalities.

Things calmed down a bit in the following two years. In 1970 there were at least thirteen non-fatal incidents, followed by at least twelve in 1971, one of which was fatal.

The year of 1972

This year saw a total of about six incidents, one of which was a major event. On 10 November three men skyjacked Southern Airways Flight 49 from Birmingham, Alabama. Their ransom demands were for $10 million, ten parachutes and a free trip to Havana. The McDonnell Douglas DC-9 with its 33 passengers and crew then began a short-haul odyssey, which took them to Jackson, Detroit, Cleveland, Toronto, Lexington, Knoxville, Chattanooga, Havana, Key West and Orlando, Florida.

At some point the hijackers wounded the co-pilot, then threatened to crash the plane into the Oak Ridge nuclear installation in Orlando. While the plane was leaving Orlando, the FBI shot the tyres on the landing gear. The plane finally made a landing on a foam-covered runway in Havana. The hijackers then surrendered.

There was one isolated instance in 1974 before the worst of all of the series of skyjackings.

6 October 1976

Authorities in Cuba believed that the saboteurs behind the bombing of Cubana de Aviación Flight 455 in 1976 had connections to Luis Posada Carriles, an anti-communist, anti-Castro paramilitary who had CIA connections, was living in exile in Venezuela and worked as an advisor for their secret service. Although Carriles admitted roles in planting bombs in hotels and was even tried in Panama in 2000 for an assassination attempt on Castro, he always denied having anything to do with the murder of the 73 people onboard CU 455.

Whatever the truth, CU 455 was scheduled to make several short trips in the Caribbean connecting Guyana to Havana. Nine minutes after taking off from Seawell Airport, Barbados—CU 455 was flying at 18 000 feet (5500 metres) when Captain Wilfredo Perez radioed: 'We have an explosion onboard. We are descending immediately! … We have fire onboard! We are requesting immediate landing! We have a total emergency!' A bomb had exploded in the Douglas DC-8's rear lavatory.

Perez immediately headed back towards Seawell for an emergency landing. Minutes later a second bomb went off. Realising that there was no hope, Perez knew that if he continued his approach, the plane might crash on a beach or into surrounding buildings. He chose to bring the plane down in the sea, 8 kilometres

TWA captain John Testrake talks to the media from his hijacked plane on 19 June 1985 at Beirut airport, at gunpoint. Hijackers commandeered the Boeing 727 on 14 June, with 153 passengers, after taking off from Athens and demanded the release of 700 prisoners from Israeli jails and the withdrawal of Israeli troops from Lebanon. After killing a US Marine, they freed all passengers 30 June.

(5 miles) west of Bridgetown. All of the 25 crew and 48 passengers died in the crash.

Two Venezuelans, Freddy Lugo and Hernan Ricardo Lozaro, travelling under false ID, had boarded CU 455 and checked in baggage in Trinidad, but had left the plane in Barbados. Trinidad authorities arrested them within hours of the disaster. They confessed to acting under the orders of Luis Posada Carriles and implicated another Venezuelan, Orlando Bosch. Police arrested Bosch and Carriles in Caracas on 14 October. The authorities of the countries concerned—Trinidad, Cuba, Barbados, Guyana and Venezuela—decided to try the four Venezuelan citizens in their home country for treason, but acquitted them in September 1980. The four men were then tried for aggravated homicide and treason in a civilian court.

On 8 August 1985 Venezuelan Judge Alberto Perez Marcano sentenced Lugo and Lozaro to twenty years in prison. The court finally acquitted Bosch on technicalities and he was finally released in 1987. He had served eleven years in prison.

Carriles made two attempts to escape before succeeding a few weeks before the Lugo and Lozaro judgement. He fled first to Panama, before finally going to America. In September 2005 a US court ruled that he could not be deported to Cuba or Venezuela because he might face torture in those countries. As of August 2006 America is holding him on charges of illegal entry into the country. The Cuban government continues to call for Posada to be brought to justice.

FAMOUS CASUALTIES

The entire Cuban National Fencing Team, many of whom were teenagers, were on the flight. They had just won all the gold medals at the Central American and Caribbean Championships.

6 January 1960

Flight: National Airlines Flight 2511
Model: Douglas DC-6B
Fatalities: 34/34 (29 passengers and 5 crew)
Principle cause: sabotage

When National Airlines Flight 2511 was en route from Idlewild International Airport (later JFK) in New York to Miami, suicidal passenger Julian Frank was presumably sitting in row seven of the DC-6B. He had just insured himself for $1 million. At 02:38 he activated a battery-detonated dynamite bomb either on his lap or under his seat. Most of the wreckage fell near Wilmington, North Carolina.

Only two months before, on 16 November 1959, National Airlines Flight 967 en route from Tampa to New Orleans also suffered a catastrophic explosion. In that case the Douglas DC-7B crashed into the Gulf of Mexico south-east of New Orleans, killing all of the 36 passengers and six crew. There is still some speculation that the two explosions were connected, but a lack of physical evidence in the earlier case, and the possibility that Julian Frank and the rest of the people on NA 2511 were unwitting victims, meant that in both cases authorities never identified nor charged anyone in connection to the bombings.

22 May 1962

Flight: Continental Airlines Flight 11
Model: Boeing 707-124
Registration: N70775
Fatalities: 45/45 (37 passengers and 8 crew)
Principle cause: sabotage

Thomas G Doty had recently been arrested for armed robbery and was due to attend hearings, when he bought sticks of dynamite, put together a home-made bomb and booked a ticket on Continental Airlines Flight 11, from Chicago to Kansas City. He was already insured for $150 000 and bought a further $150 000 of insurance (worth $2.5 million in today's terms) at the airport, before boarding the Boeing 707, which left O'Hare Airport at 20:35. Around 50 minutes later Doty put his bomb in the used towel bin of the right rear lavatory. At 21:17 the bomb exploded, completely separating the empennage (tail assembly) and causing the plane to crash 10 kilometres (6 miles) north-north-west of Unionville, Montana. People as far away as Cincinnati saw and heard the explosion. Parts of the empennage and the left wing fell some 10 kilometres away from the rest of the wreckage. By the time rescuers reached the scene, there was only one survivor, but Takehiko Nakano, aged 27, died of his injuries soon after.

A TRAGIC MILESTONE

Flight 2511 was the first confirmed sabotage of a commercial jet aeroplane.

By the 1960s pressurised cabins and jet engines had made air travel considerably more comfortable—even for hijackers.

7 May 1964

Flight: Pacific Airlines Flight 773
Model: Fairchild F-27A
Fatalities: 44/44 (41 passengers and 3 crew)
Principle cause: murder of the flight crew

Twenty-seven-year-old Francisco Gonzales was a member of the 1960 Philippines Olympic yachting team, but by 1964 he had bought insurance policies totalling over $100 000 (about $750 000 in today's terms) prior to boarding Pacific Airlines Flight 773. He told passengers that he was going to kill himself, and at 06:48 he went through with it by shooting the flight crew. The last transmission came from the first officer: 'Skipper's shot. We've been shot. I was trying to help.'

The plane went into a steep dive and a minute later hit the ground near San Ramon, California, 42 kilometres (26 miles) west of San Francisco. In the wreckage investigators found Gonzales's pistol still containing the six empty cartridges from the bullets he'd fired.

1 September 1983

Flight: Korean Air Lines Flight 007
Model: Boeing 747-230B
Registration: HL7442
Fatalities: 269/269 (246 passengers and 23 crew)
Principle cause: missile attack

One of the most famous casualties of the Cold War, Korean Air Lines Flight 007 had originated in New York and arrived at Anchorage, Alaska, at 03:00. Two hours later it was bound for Kimpo International Airport in

FROM THE COCKPIT

18:26:02 (Ambient sound in cockpit of explosion.)
18:26:06 Captain: What happened?
18:26:08 Co-pilot: What?
18:26:10 Captain: Retard throttles.
18:26:11 Co-pilot: Engines normal.
18:26:14 Captain: Landing gear.
18:26:15 (Ambient cabin sound of altitude warning alert.)
18:26:17 Captain: Landing gear.
18:26:18 (Ambient cabin sound of altitude deviation warning alert.)
18:26:21 (Ambient cabin sound of autopilot disconnecting warning alert.)
18:26:22 Captain: Altitude is going up. Altitude is going up. Speed brake is coming out.
18:26:26 Co-pilot: What? What? ...
18:27:01 Automatic Public Address Recording (APAR): Put the mask over your nose and mouth and adjust the headband.
18:27:02 Tokyo Air Traffic Control: Tokyo. Korean Air 007.
18:27:04 Co-pilot: Roger. Korean Air 007. Ah. We (are experiencing)...

18:27:08 APAR (in Japanese): Put the mask over your nose and mouth and adjust the headband.
18:27:09 Flight Engineer: All compression ...
18:27:10 Captain: Rapid decompression. Descent to one zero thousand.
18:27:15 APAR: Attention. Emergency descent. Attention. Emergency descent.
18:27:20 Flight Engineer: Now. We have to set this ...
18:27:21 Tokyo Air Traffic Control: Korean Air 007. Radio check on one zero zero four eight.
18:27:23 APAR (in Japanese): Attention. Emergency descent.
18:27:23 Flight Engineer: Speed. Stand by. Stand by. Stand by. Stand by. Set.
18:27:27 APAR: Put out your cigarette. This is an emergency descent. Put out your cigarette. This is an emergency descent. (In Japanese) Put out your cigarette. This is an emergency descent. (In English) Put the mask over your nose and mouth and adjust the headband. Put the mask over your nose and mouth and adjust ...

Nine years after the downing of KAL Flight 007, Russian President Boris Yeltsin—in a gesture of conciliation—hands over the black box recorders of the doomed plane to Korean President Roh Tae-Woo.

Seoul, Korea. Using beacon navigation the flight was supposed to have flown over Japan's north island of Hokkaido, hug the eastern coast while moving southwards and fly eventually over Honshu. Instead, far north of Hokkaido, the jumbo jet veered off course, approaching the Kamchatka Peninsula—a wilderness in the far east of Siberia.

Meanwhile, totally unrelated to the Korean Air Lines flight, a US Boeing RC-135 intelligence plane was in the sky east of Kamchatka. Alerted, Soviet defence forces scrambled six MiG-23 fighters. They may have mistaken the Korean Air Lines jet for the RC-135, because the fighters returned to base as soon as the passenger jet left Soviet airspace.

Four hours into its flight, Korean Air Lines 007 was over the Sea of Okhotsk about 300 kilometres (185 miles) off course, north of Hokkaido. Soviet defence, alerted again, scrambled one Sukhoi Su-15 Fighter at 17:42 and another at 17:54. The supersonic fighters trailed the Boeing for about 100 kilometres (60 miles). KAL 007 re-entered Soviet airspace at 18:16. Eight minutes later Soviet command ordered the plane shot down, even though it would have left Soviet airspace in two minutes. At 18:26, from a distance of about 3 kilometres (2 miles), Lt Colonel Osipovich launched one heat-seeking and one radar-guided air-to-air missile. Both hit the jumbo at a speed of over 2500 kilometres per hour (1550 miles per hour).

Although the damage initially caused the aircraft to climb, it was doomed. Japanese fishermen in the area reported seeing an orange-red fireball in the sky, which lasted about six seconds. The plane began a long descent, spiralling in ever-widening circles, as the crew struggled to keep her in the air. The cabin would have already blacked out, plunging the passengers into darkness, while the blast of explosive decompression

A Royal Jordanian Alia Boeing 727 burns at Beirut airport after it was blown up by Lebanese terrorists who hijacked the aircraft on 11 June 1985. The hijackers, who released all passengers before exploding the jet-liner, demanded the evacuation of Palestinian fighters from besieged refugee camps in Beirut to the Arab countries bordering Lebanon.

and the g-forces of the fall would have turned the plane into a freezing whirlpool of noise and confusion. It took Korean Air Lines Flight 007 four minutes to reach 16 000 feet (4880 metres) and a further eight minutes to reach 2000 feet (610 metres). After one more minute, at 18:39, the plane nosedived into the Sea of Japan.

The day after the crash, President Ronald Reagan spoke on national television, describing the attack in these words: 'It was an act of barbarism, born of a society which wantonly disregards individual rights and the value of human life and seeks constantly to expand and dominate other nations.' There were many conspiracy theories that abounded in the aftermath. The general consensus now seems to be that the tragedy was an accident born of both crew navigation error and military paranoia.

Weeks after the crash a Qantas employee told me that it 'didn't surprise' her that this had happened. It was 'well known in the industry' that Korean Air Lines pilots at the time were somewhat 'lax' in their general procedures and attitudes. Almost four months after Korean Air Lines 007, the disoriented crew of Korean Air Lines Flight 084, a cargo McDonnell Douglas DC-10-30, were taxiing at Anchorage Airport in fog, when they

failed 'to follow accepted procedures', taxied onto the wrong runway and collided with South Central Air Flight 59, a Piper Pa-31, which was on the right runway. The nine Piper passengers were injured, but, luckily, no one was killed. Whatever the truth about the competence of Korean Air Lines pilots in the 1980s, the company seems to have had more than its fair share of bad luck.

Yet this wasn't the last occasion that a plane would go down in such circumstances. Only the next time, it was the Americans doing the shooting.

11 November 1983

Flight: TAAG Angola Airlines
Model: Boeing 737-2M2
Registration: D2-TBN
Fatalities: 130/130 (126 passengers and 4 crew)
Principle cause: missile attack (suspected)

At 15:20, on 11 November 1983, a TAAG Angola Airlines 737 took off from Lubango Airport, Angola, bound for Luanda. The plane had been in the air for only a few seconds and reached an altitude of only 200 feet (60 metres), when it started to descend and turn left. The port wingtip struck the ground and the plane crashed and burst into flames, only 800 metres (0.5 miles) from the end of the runway. Officially, the Angolan authorities claimed that the crash was due to a 'technical failure'. Unofficially, rebels belonging to the National Union for the Total Independence of Angola (UNITA) claimed that they shot the plane down with a SAM missile. According to some estimates, from 1975 until 2002 the protracted civil war in Angola claimed about a million lives.

FAMOUS CASUALTIES

Conservative American Congressman Lawrence McDonald died on the flight. He hadn't meant to be on it, having missed his original flight to Seoul by three minutes at New York's JFK Airport two days earlier. McDonald's widow claimed that communists had 'murdered' her right-wing husband.

23 June 1985

Flight: Air India Flight 182
Model: Boeing 747-237B
Registration: VT-EFO, the *Emperor Kanishka*
Fatalities: 329/329 (307 passengers and 22 crew)
Principle cause: sabotage

AI 182 originated in Toronto, Canada, where operators attached a spare fifth engine to the 747's left wing, to be carried back to India for repairs. At Toronto the *Emperor Kanishka* also took on passengers and luggage. Significantly, this included luggage from Canadian Pacific Flight 60. The 747 finally left Toronto one hour and 40 minutes late, at 19:15 on 22 June, and proceeded to Montreal-Mirabel International Airport for a brief stopover.

The Canadian Pacific flight had originated in Vancouver, British Columbia, where a 'Mr Singh' had checked in a dark brown, rigid Samsonite suitcase with airline agent Jeanne Bakermans. Mr Singh failed to make his flight, but his luggage made the connection and handlers put it in the front storage section. The *Emperor Kanishka* then left Montreal for a transatlantic flight, bound for London.

It never made it.

At 07:14, at an altitude of 31 000 feet (9450 metres), while 176 kilometres (110 miles) west of Cork, Ireland, the 747 vanished from radar screens. The Transportation Safety Board of Canada later determined that a massive explosion ruptured the forward and aft cargo compartments and that the section of the aircraft behind the wings separated from the front section. The fragments of the plane, its cargo and its passengers fell into the Irish Sea into waters 2000 metres (6560 feet) deep.

Of the 329 victims, 280 were Canadian citizens, most of them of Indian descent and many of them Sikhs. Rescuers ultimately found only 131 bodies. The massive decompression had thrown some of the

passengers out of the jet long before it hit the water. For some, the forces that they encountered had ripped the clothes from their bodies. Of the 82 children on the flight, one body showed signs of drowning. The implication was terrible—some of the passengers might have survived the 10-kilometre fall only to drown while they were unconscious.

Across the world and fifty-five minutes earlier, a bomb exploded in a suitcase while handlers were loading Air India Flight 301 at New Tokyo (now Narita) International Airport. Two baggage handlers, Hideo Asano and Hideharu Koda, died; four others were injured. Airline agent Jeanne Bakermans had also checked in this suitcase in Vancouver for an 'L Singh'. Had it exploded only a little later, AI 301 would have been in the air bound for Thailand and 177 passengers and crew would have almost certainly died.

'... some of the passengers might have survived the 10-kilometre fall only to drown while they were unconscious.'

It took authorities almost three years to track down those responsible, but in truth, the investigation had started long before the deaths of the people in the Irish Sea and at Tokyo. In early 1985 police were already tagging Taliwinder Singh Parmar, leader of the Babbar Khalsa sect, dedicated to the establishment of an independent Sikh homeland in the Indian state of Punjab. In Parmar's mind 'Khalastan' would happen by any means necessary.

Police were tapping Parmar's phone and, three weeks before the crash, had followed him to a clandestine meeting in the woods of Vancouver Island with a British citizen named Inderjit Singh Reyat, and, although they heard a loud bang in the woods, the police didn't investigate further. Then a former Canadian

The Irish Navy collect fragments of Air India Flight 182, which was brought down by a bomb planted by Sikh militants in Canada in 1985. Little was left of the *Emperor Kanishka* after its 10 000 metre fall into the North Sea.

Korean Air Lines suffered more than its fair share of accidents for 'political reasons.' If such a thing as a 'fair share' can be said to exist in the first place.

intelligence agent destroyed 150 hours of taped phone conversations—before anyone had actually listened to them—because he feared that the Canadian mounted police would fail to protect the identities of informants. The mounted police, therefore, weren't following Parmar's movements on 22 June.

In a premature move the mounted police actually arrested Parmar and Reyat a few months later, only to drop the charges on Parmar and fine Reyat $200 000 before releasing him.

It wasn't until February 1988 that British police arrested Reyat on charges relating to the Tokyo bombing. The British government extradited him to Canada, and on 10 May 1991, Reyat received a ten-year sentence on two counts of manslaughter and for explosive charges for the Tokyo bombing. At this stage he denied having had anything to do with Tokyo. Parmar died in a gun battle with Indian police in 1992. He never lived to account for his role in the bombings.

In October 2000 the Canadian mounted police arrested Sikh cleric Ajaib Singh Bagri and millionaire businessman Ripudaman Singh Malik. Police charged them with murder, attempted murder and conspiracy. It was the first time in fifteen years that anyone had been charged in relation to the Air India bombing.

Reyat's trial for the Air India massacre didn't begin until 2003—eighteen years after 329 people died. He negotiated a plea bargain. For a five-year sentence Reyat pleaded guilty to manslaughter in return for a confession and information about his co-conspirators. On 16 March 2003 Baghri and Malik were found 'not guilty' because prosecutors could not establish guilt beyond a reasonable doubt.

At the time of writing, the inquiry continues, even after more than 20 years.

29 November 1987

Flight: Korean Air Lines Flight 858
Model: Boeing 707-3B5C
Registration: HL7406
Fatalities: 115/115 (104 passengers and 11 crew)
Principle cause: sabotage

In 1987 there was, to some minds, little to distinguish between Park Chung Hee's 'Fifth Republic' military dictatorship in South Korea and the 'communist' military dictatorship of North Korea, and observers continue to be suspicious of South Korea's version of the destruction of Korean Air Lines 858. However, the findings of Park's National Security Planning Agency (NSPA) and the single confession of a saboteur, with no supporting material evidence, are all we have to go on.

Twenty-five-year-old Kim Hyun Hui, also known as 'Ok Hwa', and 70-year-old Kim Seung Il worked together as spies for the North Korean regime—frequently posing as father and daughter. Fluent in Japanese, Hui and her phoney parent travelled through Europe under false passports. In Budapest North Korean agents, acting under the direct orders of Kim Jong Il, gave the pair what was to be Hui's third and, as it would turn out, final assignment. The agents told the two Kims that the successful execution of this assignment would lead to the reunification of the two Koreas. An agent then drove them to Vienna. There, they retrieved air

A TRAGIC MILESTONE

The downing of AI 182 was the greatest mass murder in Canadian history. The investigation and trials relating to Bagri, Malik and Reyat into the worst bombing in aviation history were the most expensive in Canadian history, costing an incredible CAN$130 million, including $7 million for a specially-built, high-security courtroom and $460 000 paid by the RCMP to a witness called 'John' for his testimony.

tickets secreted in a rubbish bin in a park. They were to fly to Belgrade, Yugoslavia, then on to Baghdad.

Just after midnight on the morning of 29 November, while the Korean Air Lines 707 was refuelling at Baghdad, the North Korean agents handed the Kims a briefcase. Hui went to the bathroom, where she set the timer for the bomb in the briefcase to explode in nine hours. The Kims then boarded the 707, as 'father and daughter, Shinichi and Mayumi Hachija.' The aircraft then arrived at Abu Dhabi International Airport after a short flight. The 'Hachijas' left the plane but left behind a radio filled with 350 grams of C-4 explosive and a liquor bottle containing 700 millilitres of PLX in one of the overhead racks. They then caught another flight to Bahrain.

A favourite material among scriptwriters, C-4 (Composition 4) is a military plastic explosive with the consistency of wet clay. Handlers can mould it into any shape and it detonates at approximately 8000 metres per second (26 400 feet per second). PLX (Picatinny Liquid Explosive) is a clear, yellowish fluid, which resembles white wine. According to the CIA, the effects of using both on a plane at cruising altitude are catastrophic. They later conducted a test on a 747 using the same explosive profile; the plane broke in two. However, other experts have since cast doubts that such a small amount of explosives could cause so much damage.

Korean Air Lines Flight 858 left Abu Dhabi for Bangkok at 04:01. Precisely five hours later, while the plane was over the Andaman Sea, 122 kilometres (76 miles) north-west of Tavoy, Myanmar, the bomb detonated.

Heavy air traffic to Rome delayed the Kims at Bahrain, as the world learned of the fate of KAL 858. The South Korean NSPA alerted authorities in the Middle East. Passport checks confirmed that the 'Hachijas' were fakes, and the police arrested the pair at Bahrain while they were going through departure procedures. Incredibly, they were at the end of a three-day 'holiday' here. Kim Seung Il bit into a cyanide capsule hidden in a cigarette and killed himself. At the time he was suffering from cancer, so it may not have been a hard decision for him. The North Koreans refused to accept his remains, and he is buried in a pauper's grave in South Korea.

Hui also attempted to poison herself, but authorities found her. Under interrogation she claimed to be a Chinese orphan. After she tried to kill herself again with a guard's pistol, she was extradited to South Korea. She arrived in Seoul on 15 December 1987, wearing a 'suicide prevention mask', one day before the general election forced onto General Park. Hui was put on trial under the new government of President Roh Tae Woo.

She was sentenced to death, but while awaiting execution converted to Christianity, publicly denounced her former 'Dear Leader', and was pardoned as a victim of North Korean indoctrination. Amazingly, in an open letter to President Kim Dae Jung, the families of those killed seemed to agree with this decision, but are still demanding to know the full story and the motive for the bombing. To add to the mysteries and inconsistencies surrounding the case, to date no one has ever found the wreckage, the black box or the bodies of the victims of Korean Air Lines Flight 858.

Passenger aircraft are no match for surface-to-air missiles, as demonstrated by the shooting down of Iran Air Flight 655 by the USS *Vincennes,* in a disastrous case of mistaken identity that proved highly embarrassing to the American government.

7 December 1987

Flight: Pacific Southwest Airlines Flight 1771
Model: British Aerospace Bae-146-200
Registration: N35OPS
Fatalities: 43/43 (38 passengers and 5 crew)
Principle cause: murder of the flight crew

US Airways had only recently fired David Burke, when he decided on his rather extreme revenge. As he still had his airline ID and staff at that time could by-pass security checkpoints, he was able to carry a Magnum .44 onto PSA 1771—the same flight that his former supervisor was on. The plane was en route from Los Angeles to San Francisco, when at 16:16 Burke entered the cockpit and shot the pilots. The plane nosedived into the ground near Paso Robles, California. There were no survivors.

3 July 1988

Flight: Iran Air Flight 655
Model: Airbus A300B2-203
Registration: EPIBU
Fatalities: 290/290 (274 passengers and 16 crew)
Principle cause: missile attack

Nearly five years after Korean Air Lines Flight 007, Iran Air 655 took off from Bandar Abbas Airport, Tehran, at 10:17, 27 minutes late. The Airbus was heading for Dubai in the United Arab Emirates. It was flying in daylight, in clear skies over international waters, along its approved flight path.

The Strait of Hormuz connects the Persian Gulf and the Gulf of Oman and separates the mainland of Iran from the Arabian Peninsula. It is one of the busiest channels of water in the world, with 40 per cent of the world's international oil traffic passing through it every day. Here Captain William Rogers was in command of the guided-missile cruiser USS *Vincennes*, with a mission to protect Kuwaiti oil tankers.

In the tense climate of the day, the *Vincennes* interpreted the flight of the Airbus as that of a F-14A Tomcat on an attack run. According to records the *Vincennes* challenged the Airbus seven times using Military Air Distress (MAD) frequency, referring to the jet variously as 'Iranian Aircraft', 'Iranian Fighter', 'Iranian Tomcat' and 'Iranian F-14'. Receiving no answer the *Vincennes* tried three challenges on International Air Defence (IAD) frequency, which all aircraft in the area

were required to monitor. The ship's radar monitors misread displays and thought that the plane was descending, although it was actually climbing. When the aircraft failed to respond, the *Vincennes* launched two SM-2MR surface-to-air missiles at 10:24:43, when the plane was still 18 kilometres (11 miles) away.

Captain David Carlson, commanding the USS *Sides*, had been monitoring the situation and identified the craft as a civilian on the basis of its radar signature, but he belatedly realised that the *Vincennes* was mistaking IR 655 for a Tomcat. He warned IR 655 of the situation and the crew of the Airbus immediately changed course, but it was too late.

The explosions broke off the tail and the right wing. The plane plummeted into the Strait of Hormuz killing everyone onboard.

The crew of Iran Air 655 would not have heard the MAD challenges, as their radios weren't equipped for those frequencies. Given the short space of time involved, it's doubtful that they heard the IAD challenges, and, even if they had, they probably wouldn't have assumed that the *Vincennes* was talking to them using such terms as 'Iranian Tomcat'. After all, an Airbus is three times longer than an F-14 and it would usually be a safe assumption that the two planes couldn't be confused.

US Vice-President George Bush Snr labelled the event a wartime 'incident'. Almost a month later, referring to the 'incident' at a news conference, Bush said, 'I will never apologise for the United States of America—I don't care what the facts are.' In 1990 President Bush awarded Captain Rogers with the Legion of Merit for 'exceptionally meritorious conduct in the performance of an outstanding service'.

In 1989 the Iranian government filed a case before The International Court of Justice, and, after seven years

یادبود شهدای م

سقوط هواپیمای مسافرب

توسط ناو جنگی امریک

The destruction of Iran Air Flight 655 did nothing to improve American–Iranian relations.

Various models of the Lockheed Hercules have been in service since 1959. The plane has an excellent safety record, considering the conditions under which it's used, such as this precarious approach in Pakistan. Nevertheless, considering the enormous number of flying hours amassed, accidents are inevitable.

of dilly-dallying, the American government agreed to pay the Iranians $61 million in compensation. This equates to $300 000 per wage earner and $150 000 per non-wage earner, including the 61 children who had died. On 6 November 2003 the International Court ruled that the US Navy's actions had been unlawful.

In the words of one author writing for the American magazine *Flagpole*, 'the Flight 655 incident forever ended this country's previous confident assumption that military attacks destroying civilian airliners were in all cases outrageous, barbaric, and uncivilised.' However, at least one group of people continued to regard the Iran Air incident as barbaric and their reprisal led to the most famous twentieth-century act of air terrorism, over a little town in Scotland—Lockerbie—which, until then, hardly anyone had ever heard of.

17 August 1988

Flight: Pakistan Air Force
Model: Lockheed C-130B Hercules
Registration: 23494
Fatalities: 37/37
Principle cause: sabotage (suspected)

On 5 July 1977 General Mohammed Zia ul-Haq successfully orchestrated a bloodless military coup and imposed a mixture of martial and Islamic law over Pakistan. In 1988, his eleventh year as ruler, he was returning from a tank inspection in the small town of Bahawalpur, Punjab, when witnesses reported that the Hercules carrying the President and the US

Ambassador began to fly erratically before suddenly nosediving and exploding in midair. The exact cause is controversial. Maybe someone planted a bomb—but who? Theories have indicated the CIA, the KGB, or both, Israel's Mossad or even Pakistan's own Inter-Services Intelligence. To date, there's no evidence for any of these theories and there probably never will be.

21 December 1988

Flight: Pan American Flight 103
Model: Boeing 747-121
Registration: N739PA, the *Clipper Maid of the Seas*
Fatalities: 259/259 (243 passengers and 16 crew) and 11 on the ground, total 270

The *Clipper Maid of the Seas* had arrived from San Francisco and was waiting at London's Heathrow Airport for passengers and luggage. Many of those boarding were transfers from Frankfurt, but that night PA 103 was carrying citizens from 21 countries, almost three quarters of whom were Americans.

PA 103 took off at 18:25, without incident, on its way to Kennedy Airport, New York, at a heading of 350º, almost due north. At 19:03 Shanwick Oceanic Control in Ayrshire, Scotland, transmitted clearance to the plane but received no acknowledgement. At the same moment PA 103 disappeared from the radar screen. In its place, the radar showed multiple objects fanning downwind. Eventually pieces of the aircraft travelled as far as 130 kilometres (80 miles)—to the east coast of England.

'Eventually pieces of the aircraft travelled as far as 130 kilometres (80 miles)—to the east coast of England.'

PA 103 had exploded over Lockerbie, Dumfriesshire. Subsequent events completely took over the small town of around 4000 inhabitants. According to the official report: 'Major portions of the aircraft, including the engines … landed on the town of Lockerbie and other large parts, including the flight deck and forward fuselage section, landed in the countryside to the east of the town.' The nose section was eventually found about 4 kilometres (2.5 miles) east of Lockerbie in a farmer's field.

The aeroplane's wings had detached in one piece, joined by their central structure. They hit 13 Sherwood Crescent, at the town's southern edge, at a speed of around 500 knots (930 kilometres per hour). The impact instantly vaporised several houses and formed a crater, 47 metres (155 feet) long, which spilled over 1500 tons earth from the impact site. As it had happened early in the flight, the wings were almost full of fuel, so exploded into a fireball, which set fire to neighbouring houses. At 15 Sherwood Crescent four members of the Somerville family died. They were completely annihilated, along with the bodies of those passengers who had remained attached to the wing section. There was nothing left of them to recover.

The flight recorders were found 15 hours after the crash. The plane had been functioning perfectly and had no structural flaws. The crew were mature, experienced professionals. The weather was nothing more than light, intermittent showers—normal, even pleasant for that time of year. The recorder had just stopped at 19:02:50.

At 19:04 Dumfries Fire Brigade received a call reporting a 'huge boiler explosion'. Some Lockerbie residents thought it was an earthquake; others thought that the nearby Chaplecross Nuclear Power Plant had suffered a Chernobyl-style meltdown. One saw 'a great mass of flames', another described 'a roar like an express train', while another said that 'it seemed to rain fire'.

It took fire fighters over six hours to control the blazes. Twenty-one homes that hadn't been immediately destroyed were so badly damaged that they had to be demolished.

From the evidence found in the wreckage, air accident investigators surmised that high explosives had been set off in the forwards cargo hold, and forensic scientists at the Royal Armaments Research and Development Establishment (RARDE) later identified a radio-cassette player, which had been fitted with an improvised explosive device. Although the device had been small, the results were devastating both physically and emotionally. The residents of Lockerbie had to come to terms with their town being turned into an open-air morgue.

WORDS OF A WITNESS

In the book Survivors: Lockerbie, local resident Bunty Galloway related the following:
A boy was lying at the bottom of the steps on to the road. A young laddie with brown socks and blue trousers on. Later that evening my son-in-law asked for a blanket to cover him. I didn't know he was dead. I gave him a lamb's wool travelling rug, thinking I'd keep him warm. Two more girls were lying dead across the road, one of them bent over garden railings. It was just as though they were sleeping. The boy lay at the bottom of my stairs for days. Every time I came back to my house for clothes he was still there. 'My boy is still there,' I used to tell the waiting policeman. Eventually on Saturday I couldn't take it no more. 'You got to get my boy lifted,' I told the policeman. That night he was moved.

A small inkling of the devastation of Lockerbie. The town was, literally, never the same again after having been an open-air morgue for days.

The starboard side of the nose of the *Clipper Maid of the Seas* lying in a farmer's field four kilometres from the town of Lockerbie.

It gives us an idea of how widely the wreckage was scattered.

'Careful readers will note
that this means American and European lives are worth 33 to 66 times more
than the American Government was prepared to pay
or the victims of Iran Air Flight 655.'

With tremendous compassion and stoicism, the residents of Lockerbie rallied to help the grieving relatives who were converging on the town. They had canteens open 24 hours a day. As the authorities had already determined what had happened, the villagers washed all the clothing they could find from the flight to return to family and friends.

Britain's smallest police force, the Dumfries and Galloway Constabulary, had jurisdiction over Britain's largest-ever criminal enquiry. On 13 November 1991, almost three years and 15 000 witness statements later, they and the FBI issued indictments against two men: Libyan intelligence officer Abdel Basset Ali al-Megrahi, working undercover as the head of security for Libyan Arab Airlines, and Al Amin Khalifah Fhimah, former station manager for Libyan Arab Airlines at Luqua Airport, Malta.

In 1995 there was a bounty of over $4 million on their heads for their arrest, even though, according to the Libyan government, they were supposedly under house arrest in Tripoli.

After prolonged negotiations and under the pressure of United Nations sanctions against the country, Libyan leader Muammar Gaddafi finally gave them up for trial in neutral Netherlands in April 1999. Finally, on 31 January 2001, over twelve years after the bombing, a panel of Scottish judges at Camp Zeist convicted al-Megrahi of 270 counts of murder. Fhimah also stood trial, but was acquitted and now lives with his wife and five children in Libya. Al-Megrahi is currently serving life imprisonment in Greenock Prison in western Scotland, but is appealing his sentence.

Doubts about the verdict continue to haunt the case. A United Nations appointed observer, Dr Hans Koechler, called the verdict a 'spectacular miscarriage of justice'. Many believe that the entire nation of Libya was 'framed'. Nevertheless in 2003 *The Washington Post* reported that, in a package worth $2.7 billion, Libya had agreed to pay $10 million to each family of the Lockerbie disaster—$4 million on the lifting of UN sanctions, $4 million more on the lifting of US sanctions, and the remainder 'if Libya is removed from the State Department's list of states allegedly sponsoring terrorism.'

Careful readers will note that this means American and European lives are worth 33 to 66 times more than the American Government was prepared to pay for the victims of Iran Air Flight 655.

In 2004 President George W Bush lifted sanctions against Libya—a little later than the deadline Libya had imposed—and two years later reinstated full diplomatic relations, but it remains unclear whether Libya will pay out the final $2 million per family.

FAMOUS CASUALTIES

Among the victims of the crash were Paul Jeffreys—bass player with the band Cockney Rebel—and his new wife Rachel. Ex-Sex Pistols band member John Lydon (Johnny Rotten) and his wife Nora missed the flight because Nora hadn't packed in time. Former South African Foreign Minister Pik Botha also missed the flight.

PAN AM:
A particularly unlucky airline

Several times during its 64-year history, from 1927 to 1991, Pan American World Airways paid a high price for representing American capitalism and imperialism to terrorists. Here are some of the lowlights.

17 December 1973

This was the day that five gunmen entered the terminal at Rome Airport and started shooting. They came from the Fatah Revolutionary Council, instigated by Palestinian militant Abu Nidal and were more commonly known as the Abu Nidal Organisation (ANO).

At the time, Pan Am Flight 110, a Boeing 707-321B called the *Clipper Celestial*, was preparing to taxi for its flight to Beirut. Quick off the mark, the flight's captain, Andy Erbeck, tried to protect the plane's 117 passengers and crew by aborting the taxi and ordering everyone to take cover. As the crew attempted to evacuate the plane, the terrorists threw phosphorus bombs, blowing a huge hole in the roof of the fuselage, killing 32 passengers and injuring 28 others.

Meanwhile other ANO gunmen killed an Italian customs agent, took five Italian hostages and hijacked a Lufthansa 737. There they reunited with the *Clipper Celestial* bombers and forced the crew to take them to Beirut. Lebanon refused to allow the plane to land and the 737 touched down in Athens. There the terrorists killed a hostage, threw his body onto the runway and demanded the release of two Arab militants before

flying the plane to Damascus, then Kuwait. The terrorists were eventually captured, tried and found guilty.

11 August 1982

Less than ten years later, a Jordanian terrorist with connections to the 15 May Organisation named Mohammed Rashed and his two co-conspirators, Abu Ibrahim and Cristine Pinter, placed a bomb under a passenger seat of Pan Am Flight 830, the *Clipper Ocean Rover*. The Boeing 747-121 was en route between Tokyo and Hawaii when, 225 kilometres (140 miles) from Honolulu and at an altitude of 36 000 feet (11 000 metres), the bomb exploded killing the sixteen-year-old boy who was sitting on top of it, and injuring his parents and thirteen other passengers. The blast only caused minor damage to the plane, which made an emergency landing at Honolulu.

Authorities arrested Rashed in Greece in May 1988. Convicted of murder under the Montreal Convention (an international treaty that covers the destruction of aircraft), he served eight years of a fifteen-year sentence. The FBI rearrested Rashed in June 1998 and offered him a plea-bargaining deal. In 2006 the US District Court in Washington convicted him of conspiracy and murder charges, and sentenced him to a further seven years. The court also imposed a restitution of $116 525 to be paid to the parents of the murdered boy.

5 September 1986

Abu Nidal's group hadn't finished in 1973 and struck again in 1986. PA 73 was preparing to leave Karachi, Pakistan, to fly to Munich, then on to New York, when Jordanian Zayd Hassan Safarini and three other hijackers dressed as security guards drove a van through a security checkpoint and right up to the boarding stairs of Flight 73. They stormed the plane and rapidly took charge, firing shots from a semi-automatic weapon. Fast-acting crew members were able to warn the cockpit and the flight crew escaped through a hatch.

For the next sixteen hours the hijackers threatened to kill all of the 376 passengers and remaining crew if the flight crew did not return and fly the plane to Larnaca, Cyprus, where the hijackers wanted to free some imprisoned Palestinians.

At 10:00 Safarini shot a passenger, 29-year-old Rajesh Kumar, in the head before kicking him onto the tarmac. Towards dusk the plane's mechanisms began to fail. At around 22:00 the hijackers recited a prayer of martyrdom in Arabic and when the interior lights finally went out, they opened fire and threw grenades. In the attack some people managed to open doors and escape, falling from the wings onto the tarmac, but twenty people died and more than 120 were injured while authorities stormed the plane.

Safarini spent the next five years in Pakistan. After his release the FBI arrested him on his way to Jordan. During Safarini's trial, Pakistani journalist Masror Hausen was told by a secret message that had the hijack worked, the plan was to fill the plane with explosives and ram it into the Israeli Defence Ministry. Abu Nidal, leader of the ANO and mastermind of both attacks and possibly of the Lockerbie disaster as well, was assassinated on 16 August 2002—possibly on the orders of Saddam Hussein.

'... had the hijack worked, the plan was to fill the plane with explosives and ram it into the Israeli Defence Ministry.'

Pan Am's demise

Two years later Lockerbie was the beginning of the end for Pan Am. The airline's disastrous history, a $300 million lawsuit from the families of Lockerbie, corporate mismanagement and regulatory incompetence all caused the airline to file for bankruptcy in January 1991. Pan Am Flight 436 from Barbados to Florida was the airline's last scheduled flight. The Boeing 727 *Clipper Goodwill* touched down on 4 December 1991.

19 September 1989

Flight: Union des Transports Aériens (UTA) Flight 772
Model: McDonnell Douglas DC-10-30
Registration: N54629
Fatalities: 171/171 (156 passengers and 15 crew)

Among those who are thought to have been responsible for the demise of UTA Flight 772 are the Islamic Jihad group or the Secret Chadian Resistance. Whoever did it planted a bomb onboard the DC-10 in Brazzaville, Republic of the Congo. The device was probably hidden in baggage stowed in the cargo hold, and it exploded at 13:59, 46 minutes after the plane had taken off from N'Djamema in Chad bound for Paris. The plane plunged over 35 000 feet (10 000 metres) before crashing in the Ténéré Desert in Niger.

'... a passenger entered the cockpit. He had 7 kilograms (15 pounds) of explosives strapped to his body ...'

2 October 1990

Flights: Xiamen Airlines Flight 8301 and China Southern Airlines
Models: Boeing 737-247 and Boeing 757-21B
Registrations: B-2510 and B-2812
Fatalities: 82/102 (75 passengers and 7 crew) and 46/122 (46 passengers)

Xiamen is a small regional airline based in the Chinese coastal province of Fujian. It has the distinction of being the first privately owned commercial passenger carrier in the People's Republic of China. There was no reason to believe that it would be the target of a hijacking.

Shortly after XIA 8301 took off from Xiamen at 06:57, bound for Guangzhou, a passenger entered the cockpit. He had 7 kilograms (15 pounds) of explosives strapped to his body and demanded that he be flown to either Hong Kong or Taiwan. After the man forced the rest of the crew out of the cockpit, the captain tried to bluff the hijacker, circling for half an hour before approaching Guangzhou, but telling the man that he was nearing Hong Kong.

At 08:30, as the plane landed, there was a fight in the cockpit and the plane came down hard. The Boeing 737 clipped a parked China Southwest 707B before striking the left wing and top of a China Southern Airlines 757, which was waiting for take-off clearance. As the 737 cut through the 757, the latter plane burst into flames, but the damage to the 737 was much greater. Eighty per cent of those on the 737 were killed, compared to 40 per cent on the 757. Two other parked, but empty planes were also destroyed in the collision.

Fragments from UTA Flight 772, a DC-10, lying in the Ténéré Desert in Niger.

THE FRENCH CONNECTION

Pan Am wasn't the only airline that terrorists loved to hate. Air France has seen more than its fair share of skyjackings.

Although the politically disgruntled usually see America as symbolising all of the evils of the West, a sizable proportion of the world also has it in for France. In retrospect, the possession of so many African colonies, in particular Algeria, perhaps wasn't such a good idea. Colonialism's loss, however, is our dubious gain; the historical legacy of Air France's skyjackings includes some of the most historically important events of their kind. Two in particular stand out.

27 June 1976

On 27 June 1976 Flight 139 had originated in Tel Aviv, Israel, and made a stopover in Athens, Greece, on its way to Paris. At 12:30, not long after the Athens take-off, four terrorists took over the plane and demanded that the crew fly the Airbus to Benghazi, Libya.

The terrorists were a mixed bunch. Two were from the Popular Front for the Liberation of Palestine and the other two, Wilfried Böse and Brigitte Kuhlmann, were from the obscure German group Revolutionäre Zellen. When the plane arrived at Benghazi, it remained there for seven hours, during which time the Airbus was refuelled and a female hostage was released.

The plane then took off again, and at 03:15 on 28 June arrived at Entebbe International Airport on the shores of Lake Victoria, in Uganda. Three other hijackers loyal to pro-Palestinian Ugandan President Idi Amin joined Böse and company. The hijackers transported the hostages to an old terminal building at the airport and declared an ultimatum: either the relevant authorities release 40 Palestinians held in Israel and thirteen others held in France, Germany, Kenya and Switzerland, or they would begin killing hostages on 1 July.

The hijackers then decided to release about half of the hostages, keeping only the Israelis and Jews. Michael Bacos, Captain of Flight 139 refused to leave, saying that the passengers were his responsibility, and, in a moving gesture of solidarity, the entire crew of Flight 139 chose to stay as well. A French nun also refused to leave, but Ugandan soldiers forced her and the non-Jewish passengers onto an Air France plane, which took them back to France.

The 1 July deadline passed, and the Israeli government successfully negotiated to extend the deadline to 4 July. Meanwhile the Israeli cabinet approved a rescue mission. On 3 July four Israeli Air Force C-130 Hercules made a landing at Entebbe. Another jet equipped with medical facilities landed at Nairobi.

Twenty-nine members of the elite Israeli Defence Force disguised their entry by travelling in a black Mercedes car and Land Rovers. They aimed to look like a convoy of Idi Amin and his cronies. Two Ugandan guards, who knew that Amin had recently replaced his black Mercedes with a white one stopped the convoy, and were immediately shot dead. The Israeli unit quickly stormed the building.

Defence Force member Yoni Netenyahu had given the order that the safety of the hostages took priority over the needs of the military. He was shot and mortally

An injured hostage—who was among the passengers of a hijacked Air France plane held at Entebbe in Uganda—is evacuated by Israeli soldiers. The eight-day crisis ended when Israeli troops stormed the airport building where captives were detained.

wounded by Ugandan troops as the unit entered the terminal building—but he was the only casualty among the rescuers.

The racism of the terrorists was ultimately their undoing. Ironically, the terminal had been constructed by an Israeli firm, who still had the blue prints, which helped the commandos plan the raid. The Israeli hostages themselves had been through four wars and many had military training. The hijackers even mistakenly released one French-Jewish passenger with a military background, who was able to provide the Israelis with detailed information of the situation. As it was, the fight lasted only about 30 minutes and only three hostages died.

Rather unjustly, Air France management reprimanded Captain Bacos for having stayed behind, and they suspended him from duty for a time. Israeli Prime Minister Yitzhak Rabin became hugely popular for the success of the incident, and some have cynically observed that Prime Minister Benjamin Netenyahu's later political success was partly attributable to the heroic sacrifice of his elder brother Yoni.

A hijacked Air France Airbus A-300 sits on the tarmac at Marseille's Marignane airport, 26 December 1994. Algerian fundamentalists held 172 passengers and crew hostage.

26 December 1994

On the day before Christmas 1994 Algeria was in a state of civil war. Air France Flight 8969 was at Algiers, about to depart to Paris, when a group of uniformed, armed men boarded the plane claiming to be airport security carrying out a passport check. However, the Algerian Special Forces, known as Ninjas, suspected the activity onboard the plane wasn't quite right and began to approach the Airbus. One of the armed men inside noticed the Ninjas and immediately said 'Talud' (infidels). The men abandoned their pretence, drew their weapons and shouted, 'We are not police. We are Mujahideen!'

Abdul Abdullah Yahia and three others from the Group Islamique Armé (GIA) now had 220 passengers and twelve crew members under their control. 'God has chosen us to die and for you to die with us,' and 'There is nothing to fear. God is awaiting us all in his heavenly paradise,' were just a couple of the more winning phrases Yahia shouted, as his cohorts made all of the women onboard put on headdresses and forced all of the passengers to put their personal belongings into a black plastic bag.

The Ninjas meanwhile had surrounded the plane and a stalemate ensued, during which things became very weird. Two stewardesses faked a conversion to Islam just to keep the peace. One terrorist grabbed air steward Christophe Morin on the back of the neck and kissed his forehead. Christophe then asked the terrorist, 'If you decide to kill me … I want to be able to see your face.' To which the terrorist replied, 'Don't worry. If I decide to kill you, you'll be a martyr and go straight to paradise.'

At around 14:00 the terrorists claimed their first victim. During their 'check' they had identified an Algerian policeman. They took him to the front of the plane, shot him in the back of the head and pushed his body down the front stairs as a message to the Algerian government that this would be the first of many unless they met the terrorists' demands. They wanted the staircases and wheel chocks removed so that the plane could take off to Paris.

A NOT-SO-TRAGIC MILESTONE

Known variously as Operation Entebbe, the Entebbe Incident, Operation Thunderbolt, Operation Thunderball and Operation Yonatan, the recapture of Air France Flight 139 remains one of the most successful counter-terrorist operations ever.

Philippe Legorjus was the Chief Security Consultant for Air France and a former team leader of France's elite counter-terrorism unit, the Groupe d'Intervention de la Gendarmerie Nationale (GIGN). Legorjus convened a crisis team in Paris, and the French decided that it was best to let the plane fly to France, where the GIGN could act. The French Prime Minister Édouard Balladur would allow this only if the terrorists let the women and children go. They agreed, but after 63 of the passengers had been released, the Ninja commander reneged on the deal and refused to remove the stairs and chocks. Enraged, Yahia ordered the killing of a Vietnamese diplomat and pushed the body onto the tarmac.

The stalemate continued throughout the night and the next day. In one particularly bad move, the Algerians coerced Yahia's mother to talk to her son. Enraged, he screamed at her, 'I love you, but I love God more!' before threatening to kill another hostage. At 22:00 on Christmas Day they killed French embassy chef Yannick Bernier and threatened to kill one passenger every half hour until they had their way. By now Balladur was holding the Algerian government responsible for mishandling the negotiations and for the death of a French citizen.

Thirty-nine hours after the siege started, the Algerians finally let the plane leave. Throughout this time the plane's generators had been running on full power, using up 4 tons of fuel per day, and there was now only enough fuel to reach Marseilles. Captain Denis Favier of the GIGN now hatched a plan to storm the plane in the southern French city, while the Captain of 8969 extracted a promise from Yahia that he wouldn't blow the plane up en route.

The French government now received information that the terrorists were planning to use the plane as a missile to destroy the Eiffel Tower. This was confirmed when, after arriving at Marseille, Yahia demanded 27 tons of fuel rather than the 9 tons that they actually needed to fly to Paris. The French had no intention of allowing the plane to leave Marseille. When Yahia demanded a press conference, the French agreed on the condition that the terrorists moved the passengers to the back of the plane and leave the front of the plane empty for the press to attend. Yahia agreed and played right into the hands of the GIGN. In the early winter dusk, 30 GIGN counter-terrorist specialists stormed the plane, cornering Yahia, two other terrorists and the two pilots in the cockpit. Co-pilot Jean-Paul Bauderrie threw himself out of the cockpit window and managed to stagger away, allowing a sniper to pick off the terrorists while the pilot, Captain Delain, crouched low, shielded by the controls. Soon Yahia and the other two terrorists were dead.

In the melee, eleven commandos, thirteen passengers and three members of the crew had been injured and the plane was damaged beyond repair. The fourth terrorist managed to hold out for twenty minutes before the GIGN shot him dead.

The whole event had lasted 54 hours.

OTHER AIR FRANCE ATTACKS

18 October 1973: a female hijacker held 110 people in Marseille demanding that all traffic in France be stopped for 24 hours. She released the hostages, but was killed when authorities stormed the plane.

28 August 1976: eight weeks after Entebbe, a Vietnamese man held a plane carrying twenty in Ho Chi Minh City. He released the passengers and crew, but destroyed the plane and killed himself with two hand grenades.

12 August 1977: a lone man attempted to hijack an Air France Airbus A300 near Benghazi in Libya, but police successfully stormed the plane and arrested him.

31 August 1983: four people held 111 passengers and crew during a four-day marathon from Geneva to Sicily, Damascus and finally Tehran, before finally surrendering with no loss of life.

7 March 1984: one hijacker tried to make the crew of an Air France Boeing 727-228 in Geneva take him to Libya, before he was arrested without casualties.

31 July 1984: three skyjackers commandeered a Boeing 737-228 to take them to Tehran. When they arrived after three days, they tried to blow up the plane, but failed. The authorities managed to rescue the 64 occupants.

10 December 1993: Libya was once again the destination-of-choice for a man in Paris when he hijacked an Air France A320, but authorities stormed the plane and arrested the hijacker in Nice before any of the 129 occupants were hurt.

The problem with national airlines is that political dissidents tend to see them as vulnerable, high media profile symbols of the politics of the nation itself.

23 November 1996

Flight: Ethiopian Airlines Flight 961
Model: Boeing 767-260ER
Registration: ET-AIZ
Fatalities: 125/175 (119 passengers and 6 crew)
Principle cause: skyjacking and forced ditching

Shortly after entering Kenyan airspace, heading from Addis Ababa to Nairobi, three drunken and escaped prisoners on Flight 961 skyjacked the 767 and demanded to be taken to Australia. They refused to believe the pilot when he told them that the plane would run out of fuel.

When it became apparent that the tanks were, in fact, running dry, the skyjackers tried to fly the plane themselves. The plane lost altitude, until the left

'... three drunken and escaped prisoners skyjacked the 767 and demanded to be taken to Australia. They refused to believe the pilot when he told them that the plane would run out of fuel.

engine and wingtip touched water, flipping the plane over, before it broke into several large pieces. The plane ditched in shallow water only 150 metres (500 feet) off the shore of Le Galawa Beach on the Island of Grand Comoros.

Although many people survived the initial crash, some had inflated their life jackets while still in the aircraft. This prevented them from swimming down and out of the flooded cabin and most of them drowned.

With a fleet of around 30 aircraft, of which only a very few are large passenger carriers like this L-1011, the loss of EA Flight 961 represented a considerable financial loss to Ethiopian Airlines and an incalculable human tragedy.

9/11:
A turning point in history

For a time, 'What were you doing on September 11?' was one of those questions that many people asked. The attack on the World Trade Center in New York, on September 11, 2001—and it's always September 11 or 9/11, no matter where you come from—became as much of a defining moment as the assassination of John F Kennedy had been a generation before.

Unravelling the events of 9/11 is like peeling an onion. The deeper you go, the more layers there are. Hundreds of thousands of words have been devoted to that day. There is space here to give only the briefest of accounts. It is a story that we are free to interpret in so many ways that we are still grappling with what it means, and where it will all end. To investigate further, have a look at some of the suggestions in Further Reading.

Although the plan had been set in motion for many months before the attack, the skyjacking actually began at 08:14 on American Airlines Flight 11—a Boeing 767-223, registration N334AA—with the terrorists claiming that they had a bomb. AA 11 stopped transmitting its transponder signal at 08:20 and, while flying over New York State, made a 100 degrees turn bearing south towards New York City. The five terrorists had taken control by telling the passengers that there was a medical emergency in first class. One of the terrorists probably stabbed Pilot John Ogonowski and First Officer Thomas McGuiness in order to take over the cockpit.

United Airlines Flight 175— a Boeing B-767-222, registration N612UA—left Logan Airport in Boston 14 minutes late, at 08:14. Twenty-three minutes later air traffic control asked UA 175 if they could see AA 11. They said yes, and controllers asked them to keep a distance as someone had already skyjacked it. No one imagined that, seven minutes later, UA 175 would suffer the same fate.

United Airlines Flight 77—a Boeing B-757-223, registration N644AA—left Washington Dulles Airport in Virginia, at 08:20 bound for Los Angeles. Between 08:51 and 08:54, near the boarder of Ohio, Kentucky

and West Virginia, the terrorists took over and the plane made a 180 degrees about turn—straight back to Washington.

Meanwhile United Airlines Flight 93 —Boeing 757-222, registration N591UA—was on the tarmac at Newark International Airport waiting for the traffic to clear. It finally took off at 08:42 bound for San Francisco. The 42-minute delay meant that passengers were to some extent forewarned of what would happen. As a result the fate of United 93 was profoundly different from that of the other planes.

At 08:46 terrorist Mohamed Atta rammed AA 11 into the North Tower of the World Trade Center, between floors 93 and 98 of the 110-storey building. The impact shook the building and the tower swayed alarmingly. The crash destroyed five floors instantly and trapped over 1300 people who would soon have no hope of escaping. The only way out was down, but people unsure of what to do wasted time waiting in confusion while the fire was quickly making the route down impassable. People in the already busy 'Windows on the World' Restaurant, on floors 106 and 107, continued to make calls, hoping to be rescued, until 09:40 when the phone lines went dead.

At 9:03 terrorist Marwan al Shehhi crashed UA 175 into the South Tower of the World Trade Center, between floors 78 and 84. Amazingly, some people were being told at this stage that everything was still OK and were encouraged to go back to their offices. However, most of those in the South Tower now knew for certain that this was no accident. Over 6000 people were evacuated from here over the next 53 minutes, and, with only one stairway clear, it took some people all this time to walk down more than 100 floors.

At 9:25, Ben Sliney, the National Operations Manager of the Federal Aviation Authority, on his first day in his new job, stopped all take-offs around the country—an unprecedented move that cost billions, but unavoidable.

'People in the already busy Windows on the World restaurant, on floors 106 and 107, continued to make calls, hoping to be rescued, until 09:40 when the phone lines went dead.'

Unfortunately, it was already too late for United 93. At 09:28 flight controllers heard screams from the radio in the plane's cockpit. Terrorist Ziad Jarrah, mistaking the radio for the internal public address system, pretended to be the captain and told the passengers to remain seated as there was a bomb onboard. It was the last transmission that United 93 ever made.

By 09:37 UA Flight 77 was flying just 20 feet (6 metres) above the ground, hitting streetlights, a steam vault, a power generator and the Pentagon's helipad, before Hani Hanjour managed to ram the 757 into the Pentagon. At 09:56 the upper floors of the damaged section of the building collapsed.

Just eight minutes later Ben Sliney knew he had no time to consult with FAA officials in Washington and told his staff, 'Order everyone to land, regardless of destination!'

At 09:59, the South Tower of the World Trade Center fell in a surprisingly neat collapse, almost like a controlled demolition. The one-acre broad floors fell into each other with such impact that practically everything sandwiched between them—furniture, equipment, carpet, people— pulverised into a fine grey dust that would cover lower Manhattan in a layer several centimetres thick. Witnesses reported a black so thick that they couldn't see their hands in front of their faces.

As the South Tower collapsed, the passengers of United 93 were fighting for their lives. Many had been able to sneak telephone calls to their friends and relatives, and, as a result, we know more about what happened on that particular flight than any other. The passengers also learned of the fate of the other aeroplanes. Passenger Todd Beamer told the United Airlines customer-service supervisor over the phone that he and the other passengers were planning to use a service trolley to ram the cockpit door and allow them to retake the controls. His last words on the phone were probably 'Roll it'— referring to the trolley—but they were widely reported as 'Let's roll'. This phrase became a rallying cry for American troops fighting Al Qaeda in Afghanistan.

The last thing heard on the voice recorder is the crash of metal and crockery and screaming followed by silence. There is no evidence that the passengers made it to into the cockpit, but the terrorists were somehow thwarted. UA 93 hit the field north of Shanksville, Pennsylvania, at 10:03 killing all onboard.

At this point fire fighter Captain Jay Jonas and his colleagues of Ladder Six company were in the North Tower. On reaching the twentieth floor, he made the decision to leave the building, as there was no hope of rescuing anyone any higher up. He then encountered Josephine Harris, a 'heavy-set' 59-year-old, who had already made it down 50 floors on her own, but was exhausted and struggling. Without question the men started to help her. By the fourth floor Josephine Harris's legs gave way. Moments later the collapse started.

Ladder Six company and Josephine were eleven of only fourteen people who survived the collapse of the North Tower. They happened to end up in the eye of the storm, a pocket where everything above and below them was crushed—a small island of safety amidst 1.5 million tons of debris.

On 11 December 2001 Zacharias Moussaoui, the '20th terrorist' was indicted for his part in the conspiracy and was sentenced to six consecutive life terms. His parting words to the court were, 'America, you lost and I won.'

WORDS OF A WITNESS

Captain Jay Jonas recalls:
There was also this very loud sound of twisting steel ... These massive steel beams and girders were just being twisted around our heads just like they were twist ties on a loaf of bread. And a very loud, like a steel screeching sound, almost like a lot of trains coming into a subway station at the same time and all of them hitting their brakes at the same time.

JUST THE FIGURES

3015 were killed in the attacks.

24 people are still missing and presumed dead.

20% of Americans knew someone who was hurt or killed.

1609 people lost a spouse or a partner.

3051 children lost a parent.

60 companies lost staff.

146 000 people lost their jobs in New York alone as a result of the attacks.

36 000 units of blood were donated to the New York Blood Centre.

$970 million was spent on the emergency by the Federal Government.

$1.4 billion was donated to victims and families —a quarter went to the families of the NYPD and NYFD.

$40.2 billion's worth of related insurance claims were made.

$25 million is the US Government price tag on Osama bin Laden.

Estimates for the total cost of 9/11 to Al Khaida: $250 000 to $1 million.

On 1 March, 2003, Pakistani authorities working with the FBI, apprehended Khalid Shaikh Mohammed, the alleged 9/11 mastermind, in a raid a Pakistan. According to the BBC, he is now in custody in Guantanamo Bay. As of November, 2006, he has not yet been charged in connection to the attacks.

The new World Trade Centre building and complex—to be called Freedom Tower—is due for completion in 2011. It will have 102 storeys with a total of 2 600 000 square feet (241 500 square metres) of floor space and be symbolically 1776 feet tall (541 metres).

The damage to the Pentagon is surprisingly light considering a fuel-laden Boeing 757 slammed into it at full speed..

WITH IMPROVEMENTS IN TECHNOLOGY and manufacturing, and the increased demands of the masses to become part of the 'jet set', flying has gradually become inexpensive and democratic. In fact, flying has never been cheaper.

PART THREE:
Disasters of the Jet Age

The propeller is by no means obsolete, but it is jets that now dominate the world of mass transit. Boeing estimates that, between 2004 and 2024, the world fleet of jet aircraft will grow to 35 300 passenger and cargo planes; and as the sheer number of flights increase, so do the opportunities for disaster. What follows therefore is necessarily a selection of the worst air accidents in history. 'Worst' because the chains of events claimed the most lives and led to the most catastrophic damage to property.

TRANSITION AND INNOVATION:
The Comets and the black box

Air Commodore Frank Whittle was serving in the British Royal Air Force in 1928 when he wrote a thesis on high-speed flight at high altitudes. Even at the early age of 21 Whittle had realised that there were upper limits to the effectiveness of propellers and the key to the future lay in innovation.

Whittle devised the first jet engines as early as 1929. He patented the jet in 1930, but he had to let the patent go five years later because he couldn't afford the £5 renewal fee (about $1700 in today's terms). Fortunately, two friends came to his rescue and, with a £2000 bank loan ($680 000 today), they incorporated Power Jets Ltd. With more help they were finally able to gain the UK Government's interest, but, because it took a while for them to invest any development money, Power Jets weren't able to produce a reliable engine until 1937.

Meanwhile, in Adolf Hitler's Germany, physicists Hans von Ohain and Ernst Heinkel weren't having any such difficulties. The Chancellor and his people almost immediately saw the military advantages of jets, so, even though Hans started in 1935, by 1939 the Germans already had the Heinkel He jet-powered dive bomber.

The English didn't come up with their first jet, the Gloucester E38/39, until 1941.

As groundbreaking as all this was, the development of the jet was too slow to make a significant impact on World War II. With the Allied victory, the German brain drain and a defeated Third Reich, the British became the dominant force in jet production. In 1946 De Havilland began the design of the world's first commercial jet passenger plane and in 1949 the De Havilland Comet became a reality.

Unfortunately for De Havilland, what started off as a grand vision turned into a series of tragedies. These would, however, lead to the development of one of the most significant air safety innovations ever.

The Comet's flights and crashes

The commercial jet age began with BOAC's purchase of a Comet DH 106 in 1951 and with the company's first paying passenger flight the following year, from London to Johannesburg. Able to fly twice as fast as any other passenger plane, the Comet was the Concorde

After their withdrawal from commercial service many Comets continued to serve as military transport and training aircraft.

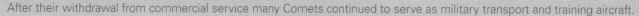

of its day, and served a similarly elite market. It was nevertheless a market large enough to justify the sale of over 50 Comets in the first year. De Havilland eventually produced no fewer than seventeen different models of the Comet.

However, a series of unfortunate crashes followed all too quickly on the heels of success. On 26 October 1952 a BOAC Flight from Campiano near Rome, Italy, failed to become airborne and some passengers received minor injuries. There were no fatalities, but the plane was a write-off. A similar crash in March 1953 (see box below: A Tragic Milestone) led to the redesign of the Comets' wings. This solved one problem but it wasn't enough.

In May 1953 a BOAC Comet flew into a heavy tropical thunderstorm and disintegrated, killing all the 37 passengers and six crew members. The enquiry determined that the cause was 'structural failure due to overstressing'. Then, in January of the following year, a BOAC flight from Singapore to London, was flying over Elba when it exploded in the air. All 29 passengers and six crew members were killed. Initially authorities suspected sabotage or an engine explosion, but the subsequent enquiry determined 'that the accident at Elba was caused by structural failure of the pressure cabin, brought about by fatigue.'

The last straw came in April 1954. Only four days after authorities had lifted the suspension on Comet flights, South African Airways Flight 201 broke up 50 kilometres (30 miles) north of Stromboli, Italy. The accident killed all fourteen passengers and seven crew. Investigators presumed that what had brought down this flight was the same as had brought down the one in January—metal fatigue—and they now set about proving it.

Finding the cause

The body of a plane experiences all sorts of stresses and strains as it takes off, flies and lands. One stress that isn't immediately obvious is that of atmospheric pressure. Over the course of its lifetime, the average jet plane frequently moves from the atmospheric pressure at sea-level to the much lower pressure of high altitudes. Jet planes travel so high that the air outside is too thin to breathe, and the aeroplane has to be pressurised from the inside so that the passengers and crew don't all asphyxiate. The end result on the body of the plane is rather like inflating and emptying a balloon, over and over again, except that in this case the skin of the balloon is made of metal. When a balloon pops, it goes quickly and explosively. The result of a high altitude depressurisation of an aeroplane is similar, but far more catastrophic.

Investigators started by piecing together what they could of the puzzle. At the Royal Aircraft Establishment at Farnborough, the British team spent six months reconstructing the recovered wreckage of the Elba crash, cataloguing every piece. Eventually they noticed a thin, black strike line in the fuselage, which could have been caused by carpet, rubber or even a coin. This confirmed that the fuselage had undergone explosive decompression.

Meanwhile investigators also set about conducting an elaborate, expensive, but incredibly important test. They submerged the fuselage of a Comet 1 into a huge tank and filled the fuselage with water to simulate the pressure of repeated climbs to and descends from 41 000 feet. The test went on for weeks with no results. Then, after the fuselage had undergone simulations of 1900 flights, it broke up. The break had started at the upper edge of a window with an instant, three-metre rupture.

A TRAGIC MILESTONE

In March of 1953 the first Canadian Pacific Comet, the Empress of Hawaii, failed to achieve the correct attitude at take-off and partially stalled. It struck a bridge just beyond the airport perimeter at Karachi, Pakistan, and crashed into a dry riverbed. All onboard died. The eleven passengers and five crew have the distinction of being the first fatalities in history of a passenger jetliner.

'Australia became the first country
in the world to require mandatory
installation of the black box on all of its aircraft.'

Not only were the windows large creating significant weak areas in the hull, they were square, and square windows undergo maximum stress at the corners. This led to the redesign of the windows, making them relatively small, with rounded edges that distribute the forces evenly.

Recording the event

Meanwhile on the other side of the world, a fuel scientist working at the Aeronautical Research Laboratories in Melbourne, Australia, was thinking about the Comet crashes. After the Dakar Comet crash, Australia's Department of Civil Aviation had formed a panel to brainstorm possible causes; in a country so large and so far away from everywhere else, Australia had a vested interest in the future of the jet. David Warren had been invited as a consultant on fuels.

Warren had only recently acquired the world's first portable wire recorder, the Minifon. Designed for the businessman to record his strokes of genius for his secretary to type up later, Warren realised that the Minifon could have other applications. It could be adapted to record voices in the cockpit, and if it was made strong enough to survive a crash, it could give investigators an idea of what had gone wrong in the event of a crash. It could also be modified to record flight control adjustments and instrument readings.

Warren presented his idea to his old supervisor, who in great bureaucratic form, suggested that he write a report. He did. He wrote several. He wrote to every aircraft authority and concern in Australia. He waited weeks for a reply. The silence was deafening. Other people in both Great Britain and America were trying to market other devices, but none had the elegant simplicity and practicality of Warren's. By 1957 Warren built his first prototype and the lack of interest remained staggering. It would not be the first time, or the last, that a brilliant Australian innovation would meet with indifference in their own country.

In the end, salvation came in the form of Robert (later Sir) Hardingham, the Secretary of the UK Air Registration Board. Hardingham was on an informal visit to Australia in 1958, when Dr Lawrence Coombes (head of the Research Laboratories) introduced him to Warren. When Hardingham saw Warren's prototype recorder, the first thing he said was, 'Now that's a damn good idea!' Hardingham put Warren on a VIP flight to the UK, where he initially encountered a lack of interest, but eventually triumphed, morally if not financially. After lengthy

An exposed 'black box' flight recorder in a military transport. To aid in recovery the boxes are not black at all, but bright orange.

wrangling The Australian Research Laboratories gave control of the patent to overseas interests, and since then many different concerns have developed the idea.

Meanwhile back in Australia, it took a tragic accident to give David Warren's idea the attention it deserved. On 10 June 1960 a Trans Australia Airlines Fokker F-27 Friendship 100 was on a scheduled passenger flight from Rockhampton to Mackay, Queensland. Mackay was enshrouded in fog, and the plane was in a holding pattern for 70 minutes. Finally given clearance to land at around 22:00, the plane crashed into the sea while on visual approach to the runway. All of the 25 passengers and four crew members died in the accident. The subsequent investigation found no material or structural defects, no fire or explosion and no evidence of unusual activity onboard. In a country with an almost perfect safety record, the crash, and the complete lack of any explanation for it, came as a deep shock. As a direct result Australia became the first country in the world to require mandatory installation of the black box on all of its aircraft.

Other countries were slower to adopt the requirement; Britain didn't adopt the idea until 1972, following the crash of a British European Airways plane into a field near the M30 motorway, killing all onboard. Investigators believed it was caused when the pilot suffered a heart attack, but without a black box, they had no way of knowing for certain. From this point on the British Government stipulated that cockpit voice recorders be fitted into all civil passenger aircraft beyond a weight of 27 tons—a stipulation now adopted internationally.

For reasons of safety and survivability flight data recorders (FDRs) and cockpit voice recorders (CVRs) have always remained separate devices and are both stored in the tail—the part of a plane that is most likely to survive because planes don't usually back into mountains.

Both machines have also undergone vast improvements over the years. They are now solid-state, sophisticated electronic devices of incredible durability. They can withstand forces of up to 3400 gs of deceleration—3400 times the force of gravity—and survive pressurised submersion in water for at least 30 days. However, the prototype remains the invention of a gentle, unassuming Australian who for years doggedly persisted in his quest to save lives—a fact that not even many Australians know.

It's often strange what survives and what doesn't. Here the landing mechanism remains intact, even after a wing has fallen off and half the cabin is gutted.

3 June 1962

Flight: Air France Flight Charter
Model: Boeing 707-328
Registration: F-BHSM, *Chateau de Sully*
Fatalities: 130/132 (all of the 122 passengers
and 8 out of 10 crew)
Principle cause: mechanical failure

Take-off is a manoeuvre that requires a lot of skill and
timing. If the plane fails to reach a particular speed by
a particular point and if the aileron settings on the wings
and tail (the trim) aren't exactly right, the plane will not
lift off the ground. In this particular crash, it seems that
something failed in this procedure.

The 707 was taking off from Paris-Orly airport for
New York when the trim servo motor failed and the
plane passed the threshold of speed and the point on
the runway for being able to take off. The pilot tried to
abort, but the plane's inertia was committed, even if
its trim wasn't. The flight crew applied the brakes while
the plane was travelling at 330 kilometres per hour
(205 miles per hour); the plane sped off the end of
runway and onto uneven grass, where the left landing
gear broke off, two engines scraped the ground and the
port wing caught fire. After sliding a further 220 metres,
the plane reached a road, the starboard landing gear
collapsed and engine number two broke off, before the
707 struck a landing light array then a house and garage.
The nose of the plane broke off before it stopped in the
Parisian suburb of Villeneuve-le-Roi, 550 metres from
the end of the runway, engulfed in flames.

20 April 1968

Flight: South African Airways Flight 228
Model: Boeing 707-344C
Registration: ZS-EUW
Fatalities: 123/128 (111 out of 116 passengers
and all of the 12 crew)
Principle cause: crew's loss of situational awareness

At 20:49 the 707 took off from Windhoek, Namibia,
in total darkness. Investigators later determined that
the crew possibly became so preoccupied with after
take-off checks that, in the absence of external visual
references, they became disoriented. Thirty seconds
after take-off the plane levelled off at only 650 feet
(200 metres) then started a rapid descent. Thirty
seconds later the 707 crashed 5327 metres (3.3 miles)
from the end of the runway at a groundspeed of 500
kilometres per hour (310 miles per hour). It was the
worst aviation accident of 1968.

3 December 1972

Flight: Spantax Charter Flight
Model: Convair CV-990-30A-5
Registration: EC-BZR
Fatalities: 155/155 (148 passengers and 7 crew)
Principle cause: crew's loss of situational awareness

Los-Rodeos on Tenerife has always been a dangerous airport on account of the mountains that surround it and its frequent bad weather. In a scenario not so dissimilar from the KLM/Pan Am disaster five years later (see page 114) the airport was shrouded in rain and fog with near zero visibility. The crew decided to take off anyway and the Convair made it to a height of 300 feet (90 metres), when the totally disoriented pilot made an ill-fated manoeuvre. At 06:45 the plane crashed at a point 15 metres to the left of the centreline of the runway and 325 metres from its end. It killed all onboard.

13 October 1972

Flight: Aeroflot Scheduled Flight
Model: Ilyushin 62
Registration: CCCP-86671
Fatalities: 174/174 (164 passengers and 10 crew)
Principle cause: navigational malfunction

The Ilyushin had travelled from Leningrad and was making its third attempt to land at Moscow during bad weather on a night when the airport's instrument landing system wasn't working. The combination of rain and darkness overwhelmed the crew and the plane crashed at the outer marker of Moskva-Sheremetyevo Airport at Krasnaya Polyana at 21:50.

Chemical foams remain the fire retardant of choice for planes where the combination of electrical wiring and highly combustible

fuels is a dangerous combination even if the plane is on the ground.

INNOCENT BYSTANDERS

Aeroplanes don't always just end the lives of passengers and crew. When they crash in populated areas they can kill people on the ground too and in rare cases—such as the 9/11 attack on the World Trade Centre —the ground casualties can be far greater than those of the aircraft's occupants.

1 February 1963

Flight: Middle Eastern Airlines Flight 265 and Turkish Air Force
Model: Vickers 754D Viscount and Douglas C-47
Registration: OD-ADE and CBK28
Fatalities: 17/17 (14 passengers and 3 crew),
3/3 (crew) and 87 on the ground, total 107
Principle cause: pilot error

The events of this accident comprise a triple tragedy —a midair collision that resulted in scores of ground deaths. Flight 265 coming from Nicosia, Cyprus, had been cleared to land at Ankara-Esenboga Airport, Turkey. The Douglas C-47 had just completed a training run at a military airfield nearby and was about to land under Visual Flight Rules. According to the official report:

The pilots of the Viscount aircraft did not see the C-47 aircraft cruising below 7000 feet on their right-hand side forward, and the Viscount, having a higher speed, caught up with the C-47 from the left rear. At the last moment the Viscount pilots saw the C-47 and tried to avoid the collision by pulling up, but they did not succeed.

The two planes intersected in midair and in clear skies at 7000 feet (2100 metres). The Viscount sliced off the C-47's left horizontal stabiliser with the lower part of its nose and its inner starboard number 3 propeller. The impact immediately ripped open the Viscount's own right fuselage. Some Viscount passengers fell out of the plane before both planes crashed in a residential area of Ankara at 15:13.

16 March 1969

Flight: Venezolana Internacional de Aviacion (VIASA) Flight 742
Model: McDonnell Douglas DC-9-32
Registration: YV-C-AVD
Fatalities: 84/84 (74 passengers and 10 crew) and 71 on the ground, total 155
Principle cause: instrument failure

Airports come in all shapes, sizes and altitudes and experience all types of weather. All aircraft models also vary in their performance parameters. The combination of these factors requires that crews program their planes so that they can take off at a particular time and place. If the crew enters the wrong information, the plane adjusts for conditions that don't actually exist. This is what happened to Flight 742.

The temperature sensors at Maracaibo Airport in Venezuela were faulty, and the crew programmed their plane based on this faulty data. At 12:00 the plane began its take-off, but left the runway too early. When it did start flying, it barely took off. The DC-9 then struck powerlines at a height of 150 feet (50 metres) and dived into the Maracaibo suburb of La Trinidad.

When aircraft crash into populated areas they pose a risk to more than just the passengers and crew.

6 March 1976

Flight: Aeroflot Flight 909
Model: Ilyushin 18E
Registration: CCCP-75408
Fatalities: 120/120 (109 passengers and 11 crew)
and 7 on the ground, total 127
Principle cause: system failure

It was 01:00 and Flight 909 was flying at an altitude of
26 000 feet (7900 metres) in complete darkness, with no
visible horizon on its way from Moscow to Yerevan. The
crew was relying entirely on their instruments when an
electrical failure disabled the compass, two main gyros
and the autopilot. The crew struggled to make sense
of their correct position and, apparently in an act of
confusion, tried some banking manoeuvres before losing
control of the plane and crashing near Verkhnyaya Khava
in Russia. All onboard died and the plane killed another
seven people on the ground.

11 October 1984

Flight: Aeroflot Flight 3352
Model: Tupolev 154B-1
Registration: CCCP-85243
Fatalities: 174/179 (169 out of 170 passengers and
5 out of 9 crew) and 4 on the ground, total 178
Principle cause: criminal negligence

Flight 3352 was touching down at Omsk, Russia, at
05:41. The weather was poor, with light rain after snow
and with clouds only 300 feet (90 metres) above the
ground. The plane touched down and was travelling at
260 kilometres per hour (160 miles per hour), when the
crew noticed that there were still two snow-clearing
vehicles on the runway. They failed to steer the plane in
time and the plane crashed, killing everyone onboard
as well as the four snow removal workers.

A flight controller had fallen asleep on the job and
failed to inform the approach controller where the snow
ploughs were.

'A flight controller had fallen asleep
on the job and failed to inform the
approach controller where the snow
ploughs were.'

3 September 1989

Flight: Cubana de Aviacion Chartered Flight
Model: Ilyushin 62M
Registration: CU-T1281
Fatalities: 126/126 (115 passengers and 11 crew)
and 45 on the ground, total 171
Principle cause: pilot error

The Ilyushin was in its initial climb from Havana Airport
when heavy rain and 30–40 kilometres per hour (19–25
miles per hour) winds got the better of it. A severe and
sudden downdraft caught the plane at a height of 150
feet (46 metres). At 19:00 it struck navigational aerials
then crashed into a residential area.

4 October 1992

Flight: El Al Flight 1862
Model: Boeing 747-258F
Registration: 4X-AXG
Fatalities: 4/4 (1 passenger and 3 crew) and 39 on
the ground, total 43
Principle cause: separation of engines from wing

Flight 1862 lifted off from Amsterdam's main airport at
17:21 bound for Tel-Aviv. It had reached 6500 feet (2000

metres), when the inner right number three engine and pylon separated from the wing in such a way as to collide with the outer right number four engine, causing the separation of that engine and pylon as well. After the crew had declared an emergency, they had to turn the plane in order to lose enough height and speed in order to touch down again. The crew had to abort the first approach when ATC realised that the 747 wasn't going to make it. On the second attempt the crew lost control of the plane, and the 747 crashed into the eleventh floor of an apartment building in the Amsterdam suburb of Bijlmermeer. The official report cited bad design and faulty inspection procedures as contributing factors in the failure of the engine pylons.

8 January 1996

Flight: African Air Cargo Flight
Model: Antonov 32B
Registration: RA-26222
Fatalities: 2/6 (crew only) and at least 225 on the ground, total 227
Principle cause: criminal negligence

It was 12:23 when the cargo plane attempted to take off from Kinshasa-N'Dolo Airport in Zaire (now the Democratic Republic of the Congo). Unconfirmed reports state that the plane was 270 kilograms (595 pounds) overloaded, that African Air did not have

The devastation caused in Amsterdam when an Israeli cargo plane crashed into an apartment building

the authority to fly the plane and had borrowed the authorisation papers of another carrier. Other reports state that the plane may have been carrying weapons for rebels for the Union for the Total Independence of Angola (UNITA). No matter what regulations African Air was bending, the Russian crew flying the plane couldn't bend the laws of physics and make the plane gain any height.

The Antonov overshot the end of the runway and crashed into Simbazikita, a crowded open-air vegetable market next to the airport. The plane ploughed a fiery 100-metre (330-foot) swathe through people, cars, trucks and wooden and tin shacks.

The disaster brought out the best and worst in people. In the chaos and hysteria, some did all that they could to help, while looters ripped off jewellery from burning, severed heads, arms and legs and ransacked ruined stalls. The four crew members managed to escape the wrecked plane, but then had to fight off a large mob of angry, murderous people bent on revenge.

The estimate of 225 ground fatalities come from reports in the English newspapers *The Weekly Mail* and *The Guardian*, which covered the trial of the pilots. However, other reports suggest that as many as 350 people in the market may have died. There were well over 500 injured, and the disaster completely overwhelmed Zaire's largest hospital, Mama Yemo, which had been designed to accommodate only 60 people. The victims were so thoroughly damaged by the plane's propellers, the fire and the melee that only 66 bodies were ever properly identified.

Both the pilot, Nicolai Kazarin, and the co-pilot Andrei Gouskov, stood trial for 225 counts of murder. Kazarin claimed that, 'the market shouldn't have been there', and both pilots blamed each other for the crash. In an outcome that exemplifies Zairian justice at the time, both pilots were convicted of homicide and were each sentenced to a maximum of only two years in prison.

Over a decade on, it is still Africa's worst ever air disaster. No aviation disaster other than 9/11 has claimed so many fatalities on the ground, and yet it remains largely forgotten by the rest of the world.

WORDS OF A WITNESS

An N'Dolo Airport employee later said that the plane 'did not even get its nose up. It was going as fast it could, but didn't even manage to take off.'

> 'The disaster brought out the best and worst ...
> some did all that they could to help,
> while looters ripped off jewellery from burning,
> severed heads, arms and legs and ransacked ruined stalls.'

16 February 1998

Flight: China Airlines Flight 676
Model: Airbus A300-622R
Registration: B-1814
Fatalities: 196/196 (182 passengers and 14 crew)
and 7 on the ground, total 203
Principle cause: pilot error

Rain and fog enshrouded Taipei Airport. Flight 676's pilots aborted their landing when they came in 1000 feet (300 metres) too high. The pilots then tried to climb again but didn't achieve the correct pitch and airspeed, and, while travelling at only 83 kilometres per hour (52 miles per hour), the Airbus stalled. At 20:09 the plane hit the ground 65 metres (200 feet) left of the end of the runway and skidded into several houses, factories, fish farms, rice paddies and warehouses before exploding. In the light of the incident, China Airlines, retrained all of its 670 pilots.

4 May 2000

Flight: Executive Airline Services (EAS) Flight 4226
Model: BAC 111 525FT
Registration: 5N-ESF
Fatalities: 71/77 (64 out of 69 passengers and 7 out of 8 crew) and 78 on the ground, total 149
Principle cause: mechanical failure

Prior to its last flight, the BAC 111 had spent 52 days on the ground because of engine troubles. Mechanics finally fitted an engine cannibalised from another 111 onto 5N-ESF only ten hours before it was due to leave for Lagos, Nigeria.

Shortly after its 13:35 take-off from Kano, Nigeria, the 22-year-old plane stalled and flew into the heavily-populated suburb of Gwammaja. In the crash the plane destroyed 23 houses, a school and sliced a mosque in half.

The Aviation Ministry of Nigeria grounded all BAC-111s pending the outcome of the investigation and banned all aircraft older than 22 years from flying in Nigerian airspace.

5 September 2005

Flight: Mandala Airlines Flight 091
Model: Boeing 737-230
Registration: PK-RIM
Fatalities: 107/122 (102 out of 117 passengers and all of the 5 crew) and 41 (suspected) on the ground, total 148 (suspected)
Principle cause: unknown

The exact causes and death toll of this accident are uncertain because of varying accounts. What we do know is that Flight 091 had just taken off from Medan-Polonia Airport in Indonesia, bound for Jakarta, when at 09:40 the plane shook violently, veered left and crashed into a residential district 460 metres (1500 feet) from the end of the runway.

13 October 1972

Flight: Uruguayan Air Force Flight 571
Model: TAMU Fairchild Hiller 227D
Registration: T571
Fatalities: 29/45 (24 out of 40 passengers and all 5 crew)
Principle cause: turbulence

Although not a numerically high fatality, this accident became famous not only because of the occupants but because of what they had to go through after the crash. The 227D had left Mendoza, Uruguay, carrying the Stella Maris Rugby Union Team to play a match in Santiago, Chile. It was flying across the Andes at an altitude of 15 000 feet (2800 metres) when the plane hit severe turbulence on the border of Chile and Argentina.

The right wingtip hit a mountain at 4200 metres (13 800 feet) and broke off. Soon after the right wing folded over the main fuselage, ripping a hole into the cabin. The remains of the wing then sheared off a vertical stabiliser on the tail. The plane clipped another peak, which broke off the left wing. The fuselage then hit the slope of an unnamed mountain at a speed of 350 kilometres per hour (217 miles per hour) and slid down a steep incline, before coming to rest at a snow bank. The mountain would later be called Cerro Seler and the specific slope the Glacier de las Lagrimas (the Glacier of Tears).

As Flight 571 crashed the plane's seats piled up together. Thirteen occupants died immediately, another five by the next morning. The 27 survivors were suffering from multiple fractures and other serious injuries and were now trapped on a remote mountain, in freezing weather, in the fragmented remnants of their former plane, without protective clothing or medicines and only a few chocolate bars, snacks and some bottles of wine for food—hardly enough to sustain them in an environment where calorific requirements are huge.

Rescue teams from Chile, Argentina and Uruguay tried in vain to find a white plane against a white background and called off the search after eight days, during which another passenger died. On their eleventh day on the mountain, the 26 survivors had found a small transistor radio and heard the news that rescuers had given up on them. One of the survivors, Gustavo (Coco) Nicholich, concluded that this was good news, 'Because it means that we're going to get out of here on our own.'

So on 24 October the boys decided that they would have to eat the bodies of their friends—a profoundly difficult decision for anyone, let alone Roman Catholics, and a decision that some could not take for several more days. Nando himself had lost his mother and sister in the crash.

Then on 29 October eight more died when an avalanche descended on them while they slept. There were now 18 survivors left.

WORDS OF A SURVIVOR

Even with stringent rationing, there wasn't enough food. In his book Miracle in the Andes, survivor Nando Parrado spoke about the famous decision that the young men were then forced to make:

... we were starving in earnest, with no hope of finding food, but our hunger soon grew so voracious that we searched anyway ... We tried to eat strips of leather torn from pieces of luggage, though we knew that the chemicals they'd been treated with would do us more harm than good. We ripped open seat cushions hoping to find straw, but found only inedible upholstery foam ... Again and again I came to the same conclusion: unless we wanted to eat the clothes we were wearing, there was nothing here but aluminum, plastic, ice, and rock.

The Andes are as beautiful as they are brutal. The top of a mountain is probably the worst place that you could be after an air crash.

The more able-bodied decided to search for anything useful they could find. One excursion found the tail, which still contained some provisions and some insulating material. Momentary hope died when, after several days of trying, they failed to make the plane's radio work from batteries that they'd also found in the tail.

The survivors now realised that they had to travel west to find help, and they all pitched in to make a sleeping bag from the insulating material. After the completion of the sleeping bag and the death of another survivor the expeditionary group comprising Nando Parrado (22), Roberto Canessa (19) and Antonio 'Tintin' Vizintin (19) made ready to go.

It was 12 December and the now 17 survivors had been on the mountain for more than two months. After three days of finding their way through the mountains, the trio were going through their rations too quickly and decided that Vizintin should return. It took him only three hours to make it back.

It took Parrado and Canessa three more days to reach the end of the snowline. On the evening of their ninth day, they came to a river where they found and called out to a group of three Chilean horsemen, but couldn't reach them or hear them clearly. The next day one of

the horsemen, Sergio Catalan, returned, tied paper and pencil to a rock and threw it across the river and Nando Parrado wrote down an explanation. Sergio then rode off to find help. Finally, several hours later, helicopters arrived. It was 21 December.

Back on the mountain, one more had died, but on 22 December the fourteen remaining survivors heard on their small transistor that Parrado and Canessa had found help. That afternoon Parrado guided the rescue team to the crash site. Because of the hazards of flying in the Andes, rescuers took two days to evacuate them off the mountain.

Initially the survivors wanted to tell their families in person how they had made it, but sensationalist reports leaked to the press. On 28 December the survivors held a press conference, while still suffering from the after-effects of altitude sickness, bone fractures, dehydration, frostbite, scurvy and malnutrition as well as the psychological trauma of cannibalism—but they told their story. In all 29 people had died. The sixteen young men who had survived were as young as 19 and mostly in their early 20s. They had held on for 72 days. Their story would be told again and again, most famously in Piers Paul Reid's *Alive* and a documentary based on Reid's book.

After over two months on a freezing mountain in the Andes, survivors of Flight 571 wave to a rescue helicopter. Suffering from altitude sickness, bone fractures, frostbite, scurvy and malnutrition, they also had to deal with the psychological trauma of cannibalism.

'The sixteen young men who had survived were as young as 19 and mostly in their early 20s. They had held on for 72 days.'

The public, though disturbed, understood the hell that they had gone through. Nando Parrado related that:

Shortly after our rescue, officials of the Catholic Church announced that according to church doctrine we had committed no sin by eating the flesh of the dead. As Roberto had argued on the mountain, they told the world that the sin would have been to allow ourselves to die. More satisfying for me was the fact that many

of the parents of the boys who died had publicly expressed their support for us, telling the world they understood and accepted what we had done to survive.

According to Nando, the press wasn't nearly so kind,

'... despite these gestures, many news reports focused on the matter of our diet, in reckless and exploitive ways. Some newspapers ran lurid headlines above grisly front-page photos.'

Nando Parrado developed a career as a professional racing car driver and later became a television personality and motivational speaker, using his experiences to help others deal with trauma.

WINDSHEAR

You don't need to go into remote regions and high mountains to encounter violent wind conditions. For a period of about twenty years between 1964 and 1985, the atmospheric phenomenon of windshear was responsible for or contributed to some 26 major civil aviation accidents in America alone.

Windshear technically refers to the difference in wind speed and direction between any two points in the atmosphere. Since jet planes travel fast, they may rapidly encounter sudden changes in wind rates and vectors. This is particularly true during thunderstorms.

The most significant hazard for an aeroplane is that rapid changes can occur near the ground where there is very little vertical room to manoeuvre. For example, as a plane is landing, there may be a strong headwind that forces the pilots to reduce engine power in order to lower airspeed. As the plane continues towards landing, the headwind may suddenly vanish just at a time when the plane needs to maintain enough altitude. If the pilot doesn't react in just the right way, there may not be enough time to power the engines up again and the plane may simply drop, or the pilot could overcompensate and come down too fast. Similar problems can occur on take-off. Either extreme at this most delicate part of the flight could be potentially catastrophic, and often was.

7 July 1980

Flight: Aeroflot Flight 4225
Model: Tupolev 154B-2
Registration: CCCP-85355
Fatalities: 163/163
Principle cause: windshear

Aeroflot Flight 4225 had reached an altitude of no more than about 500 feet (150 metres), two minutes into its flight from Kazakhstan to Ukraine, when it reached a zone of hot wind. The plane's airspeed suddenly dropped and it was simultaneously caught in a down draft. It stalled then plunged nose-down into a wheat-field, where it disintegrated, caught fire and slid until it fell into a ravine.

Aeroflot is an airline with a dubious safety record, but windshear can happen to any plane, anytime.

9 July 1982

Flight: Pan Am Flight 759

Model: Boeing 727-235

Registration: N4737

Fatalities: 145/145 (138 passengers and 7 crew) and 7 ground crew, total 152

Principle cause: windshear

The 727 had just taken off from New Orleans and had reached an altitude of 95–150 feet (29–45 metres) when it lost height and struck trees 724 metres (2376 feet) from the end of the runway. It crashed killing all onboard and demolishing six houses. The official report gave a typical account of the devastating effects of windshear: 'Contributing to the accident was the limited capability of current ground based low level windshear detection technology to provide definitive guidance for controllers and pilots for use in avoiding low level windshear encounters.'

2 August 1985

Flight: Delta Air Lines Flight 191

Model: Lockheed L-1011 TriStar

Registration: N726DA

Fatalities: 134/163 (126 out of 152 passengers and 8 out of 11 crew)

Principle cause: windshear

The crew of Flight 191 flew through a lightning-filled cumulonimbus cloud, then attempted to land at Dallas–Fort Worth, Texas, during a thunderstorm in conditions of severe low altitude windshear, for which they were unprepared and untrained.

In a matter of seconds the plane experienced variations of airspeed from 320 kilometres per hour (200 miles per hour) to 220 kilometres per hour (137 miles per hour). The L-1011 touched down 1800 metres (6300 feet) short and 110 metres (360 feet) left of the runway centreline before becoming briefly airborne again. As it fell to earth a second time, it crossed State Highway 114, struck a car and killed the driver before skidding on to the airfield. The empennage broke free from the main fuselage and it was this tail section that carried most of the survivors. Most of the plane then crashed into two water tank reservoirs before exploding into flames.

The severity and frequency of windshear-related accidents led to the development of sophisticated windshear detection systems and programs for flight crew training. In 1988 the Federal Aviation Authority mandated that all commercial aircraft flying in American airspace had to be fitted with such devices and by 1995 all relevant airlines had complied. Since then, the number of windshear related accidents have dropped from an average of more than one per year to one every ten years.

FAMOUS CASUALTIES

Onboard Delta Air Lines Flight 191 was Don Estridge, the principle engineer behind the development of the IBM PC, along with his wife Mary and many of the other IBM executives who had developed the basis for a lot of modern hardware. IBM changed its corporate travel policy on account of the accident and no more than two IBM executives can now fly on the same plane. A similar policy governs the travel arrangements of the British Royal Family.

29 December 1972

Flight: Eastern Airlines Flight 401

Model: Lockheed L-1011 TriStar

Registration: N310EA

Fatalities: 103/176 (94 out of 163 passengers and 5 of 13 crew immediate, 4 more died later from injuries)

Principle cause: controlled flight into terrain

The introduction of wide-bodied aircraft like the Lockheed L-1011 immediately raised the stakes of potential air fatalities to number in the hundreds.

'The failure of a fifty-cent light bulb and a flawed autopilot design had distracted a crew until they were seconds from crashing.'

Captain Robert Loft, First Officer Albert Stockstill and Flight Engineer Donald Repo had already flown from Tampa to New York without incident and were ready and eager to take a plane full of passengers returning from their Christmas trips to their hometown of Miami. Although there had been two non-fatal incidents involving L-1011s in the previous month, the aircrew had no reason to believe that their state-of-the-art 'Whisperliner' or any of its 250,000 parts would fail them. Eastern Airlines maintenance procedures were sound; they checked their aeroplane's circuits every three days. In seven years of operations, and with 1400 flights per day, the airline had never had an accident.

The weather that Friday was perfectly clear. As Flight 401, the plane had left New York's Kennedy Airport on time at 21:20 and it was still on schedule over two hours later when it was making its final approach to Miami and the aircrew engaged the landing gear. At this point Stockstill noticed that the green landing gear light was still unlit, seemingly indicating that the front landing gear had failed to go down. They retracted the landing gear and made a second attempt. Again, no light. Either the light was blown or the gear's hydraulics had failed. It didn't matter, as the crew could crank the landing gear manually.

Captain Loft asked for instructions. ATC Miami told him to abort landing, climb to 2000 feet (600 metres) and turn west over the Everglades. While the cockpit crew fiddled with the lights Don Repo went into the 'hell hole'—the avionics bay located beneath the cockpit. There he could look through a telescopic peephole and see if the landing gear was, in fact, down.

After 50 seconds at 2000 feet Loft asked Stockstill to put the plane on autopilot. The plane maintained a level flight path for 80 seconds while the crew continued to investigate the landing gear issue. Then the plane dropped 100 feet (30 metres), flew level for 120 seconds, then began a gradual descent. The subsequent investigation showed that there was a crucial flaw in the autopilot disengagement system. Normally if the pilot or the co-pilot apply a certain amount of pressure to the controls, the autopilot automatically switches off.

In this case it seemed to have taken a only a tiny amount of pressure to disengage the autopilot and this may have happened while Loft was turning in his seat to talk to his crew, perhaps bumping his control column with his leg or hip. Furthermore, there was no obvious signal that the autopilot had disengaged. It only took a small amount of forward pressure on the control column to nudge the plane into a relatively gentle, but constant, descent.

Over the next 70 seconds, the plane lost about 250 feet (75 metres) of altitude. This activated an alert chime from the flight engineer's workstation. Don Repo, in the hell hole, didn't hear it. Nor it seemed did any of the other crew. Fifty seconds later the plane was still going down. Only when it was time to make a turn back to the airport did Stockstill notice the altitude. It was too late. At 23:42:13 Flight 401 crashed into the Florida Everglades 30 kilometres (19 miles) from the end of Miami's Runway 9. The left wingtip, engine and landing gear touched first, then the fuselage of the plane hit the swamp. It tore about 180 tons of aircraft and people apart as the nose turned clockwise and the main body split into five large pieces. As the fuel tanks burst, flames engulfed the cabin.

The NTSB would later conclude that the front landing gear was down and locked. It always had been. The failure of a fifty-cent light bulb and a flawed autopilot design had distracted a crew until they were seconds from crashing.

The rescue was difficult. The Everglades is a huge swamp about 200 kilometres (120 miles) long and 112 kilometres (70 miles) wide, it's thigh-deep in brackish water, and teeming with insects, alligators and poisonous snakes. What greeted the rescuers when they arrived just after midnight was a mixture of random destruction and salvation. Some survivors had barely a scratch; some were as burned and twisted as much of the plane. Many, both alive and dead, had had their clothes torn off. Mercifully, the dark hid much of the sight, but could not hide the wailing or the screaming, or in fact the Christmas carols that a group of survivors had started to sing from a part of the wreckage. Survivors and bodies were taken to nearby Palmetto and Mercy Hospitals. Eventually the average legal settlement for the survivors was $250 000—about $1.5 million in today's terms.

Out of this mess, a ground engineer, Angelo Donadeo, who had been sitting in the cockpit jump-seat managed to survive and give an insight into what had happened. Although his back was broken in the crash he returned to work only three months later. He had already had one miraculous escape in his life, when he was wounded in World War II, with first, second and third-degree burns all over his body, when his ship had been hit by a kamikaze pilot.

THE GHOST OF FLIGHT 401

Over the few years following the crash of Eastern Airlines Flight 401, something strange started happening. Eastern Airline employees and passengers reported sightings of both Don Repo and Bob Loft on various flights. A typical story involved smelling the aftershave of the dead men in familiar places.

According to the Flight 401 memorial website:
'A flight's captain and two flight attendants claim to have seen and spoken to Loft before take-off and watched him vanish —an experience that left them so shaken they cancelled the flight … one female passenger made a concerned enquiry to a flight attendant regarding the quiet, unresponsive man in Eastern Airlines uniform sitting in the seat next to her, who subsequently disappeared in full view of both of them and several other passengers, leaving the woman hysterical. When later shown a sheet of photos depicting Eastern flight engineers, she identified Repo as the officer she had seen …

On another occasion, Faye Merryweather, a flight attendant, saw Repo's face looking out at her from an oven in the galley of TriStar 318. Understandably alarmed, she fetched two colleagues, one of whom, the flight engineer, had been a friend of Repo's and recognised him instantly. All three heard Repo warn them to 'Watch out for fire on this aeroplane'. Not long after, the aircraft lost an engine while taking off from Mexico City, then lost a second engine to fire moments after that. The plane landed safely.'

One of the worst places to crash an aircraft—an alligator and insect-infested swamp like the Florida everglades.

4 December 1974

Flight: Martinair Holland Flight 138
Model: Douglas DC-8-55F
Registration: PH-MBH
Fatalities: 191/191 (182 passengers and 9 crew)
Principle cause: pilot error

As they approached Colombo Airport, Sri Lanka, the flight crew of Flight 138 failed to recognise their correct height and distance relative to their position from the airport. They continued a premature descent until they crashed into Mount Anjimalai in Maskeliya at an altitude of 4355 feet (1330 metres) around 40 kilometres (25 miles) east of the capital.

A TRAGIC MILESTONE

With all 191 onboard killed Martinair Holland Flight 138 remains Sri Lanka's worst accident so far.

3 August 1975

Flight: Aila Royal Jordanian Airlines Charter Flight
Model: Boeing 707-321C
Registration: JY-AEE
Fatalities: 188/188 (181 passengers and 7 crew)
Principle cause: pilot error

At 04:25 the 707 was on its final approach to Agadir Airport in Morocco when the right wingtip and engine number 4 struck a mountain at a height of 2400 feet (730 metres). That part of the wing and the engine broke off, the pilots lost control and the plane crashed into a ravine at Immouzer, 40 kilometres (25 miles) north west of Agadir.

Over 30 years on, the Aila Royal Jordanian crash remains Morocco's worst air disaster.

1 September 1975

Flight: Interflug Flight 134
Model: Tuplolev 134
Registration: DM-SCD
Fatalities: 27/34 (24 of 28 passengers and 3 of 6 crew)
Principle cause: crew error

The radar controller failed to monitor the last 3200 metres of Flight 134 as it approached Leipzig and the plane's crew didn't check their altitude. The Tupolev flew below its glide path and struck a radio tower when it was only 6–9 feet (2–3 metres) off the ground. The collision damaged the left wing, the right engine separated and the plane crashed just short of the runway. The captain was jailed for five years and the co-pilot, navigator and radar officer for three-years each.

'Continuing their fantasy-land assumption, the crew crashed the plane into the Karatepe Mountains.'

19 September 1976

Flight: Turk Hava Yollari Scheduled Flight
Model: Boeing 727-2F2
Registration: TC-JBH
Fatalities: 154/154 (146 passengers and 8 crew)
Principle cause: navigation error

A truly bizarre case of mistaken identity. The crew of the 727 had departed Istanbul at 22:45 on their way to Antalya, on the southern coast of Turkey. Even at night this city is easy to recognise from the air because of its 4-kilometre (2.5-mile) long boulevard. At 23:11 the crew radioed ATC, reported seeing the Antalya city lights and requested an approach. In reality they were about 200 kilometres north and approaching the city of Isparta. Continuing their fantasy-land assumption, the crew crashed the plane into the Karatepe Mountains.

EVERY PILOT'S NIGHTMARE
The runway collision — Tenerife 1977

27 March 1977

Flight: Pan American World Airways Flight 1736 and
KLM Flight 4805
Model: Boeing 747-121 and Boeing 747-206B
Registration: N736PA, *Clipper Victor*, and PH-BUF,
Rhine River
Fatalities: 335/396 on Pan Am 1736 (326 out of the 380
passengers and 9 out of the 16 crew) and 248/248 on
KLM 4805 (234 passengers and 14 crew), total 583
Principle cause: collision on runway

The tragedy of errors that would culminate in the
worst civil aviation accident in history to date began
indirectly because of an act of terrorism. However,
as the black box transcripts and the subsequent
investigation revealed, there were many times
when the players could have averted catastrophe.
Unfortunately, the parties involved made the wrong
decisions. What looks like an inevitable accident
wasn't inevitable at all, but then again we have the
benefit of hindsight.

At 12:30 members of a militant Canary Islands
independence group exploded a bomb in a flower
shop at the Las Palmas passenger terminal on
Gran Canaria. The bomb injured nine people and
authorities were concerned about a threat that there
might be a second bomb, so they closed Las Palmas
and diverted all air traffic to Los Rodeos on the island
of Tenerife. Although Las Palmas reopened at 15:00,
Los Rodeos was overcrowded by then, causing
delays at the smaller airport.

The *Rhine River* left Amsterdam at 09:31. The
234 passengers on a winter holiday included three
babies and 48 children. It touched down at Tenerife at
13:38 where there were already three other planes
under the control of only two air traffic controllers.
The plane's captain was Jacob Veldhuyzen van
Zanten, KLM's most senior and experienced pilot
and principle trainer on 747s. He decided to refuel
the plane, which would take half an hour and, as he
didn't know how long Las Palmas would be closed,
he let the passengers disembark and stretch their
legs at the Tenerife terminal. Already the island was
experiencing low cloud and light rain.

Half an hour later, at 14:15, the *Clipper Victor* arrived
from New York full of senior citizens looking forward to
a twelve-day cruise. Within fifteen minutes authorities
declared Las Palmas clear, and although ATC Los
Rodeos could have put the *Clipper Victor* in a holding
pattern, they chose to divert the Pan Am Flight instead
to land at Las Palmas.

At Tenerife the *Clipper Victor* was ready to leave but
couldn't take off. Los Rodeos airport is much smaller
than Las Palmas. Only runway 30 is large enough to
accommodate a 747 take-off but the *Clipper Victor*
couldn't access the runway now because of congestion
behind it. Nor could it taxi along runway 12 to reach
runway 30 because the *Rhine River* was blocking
it. As a result Pan Am captain Victor Grubbs and his
passengers would be stranded on the runway for
hours waiting for the congestion to clear.

The remains of the day. Human investigators dwarfed by the remnants of the worst aircraft disaster in history until the 2001 attack on the World Trade Centre in New York.

A cold Spanish policeman stands guard in the graveyard that was Tenerife's international airport.

'There he is ... Look at him!
Goddam that son-of-a-bitch is coming!
Get off! Get Off! Get Off!

Finally, at 16:56, ATC cleared KLM to taxi into position for take-off. This meant taxiing and reaching the north-west point of runway 30, then turning 180 degrees so that it would be facing south-east for take-off along runway 30.

Shortly after this ATC spoke to Pan Am, telling them to taxi down the runway and leave the runway third to their left.

At this point the *Clipper Victor* was behind the *Rhine River* on runway 30.

Tenerife ATC then began to give instructions to the *Rhine River*, telling them to turn 180 degrees at the end of the runway and wait for instructions. Unfortunately, the weather was deteriorating and a fog was descending over the airport. In the air traffic control tower, the controllers couldn't see the planes and there was no ground radar. The planes couldn't even see each other.

At this point the Pan Am crew, who were unfamiliar with the airport, were showing signs of confusion about just where to turn to be in the right position. They knew that the Dutch were in front of them and that they had to taxi off the runway so that the prior flight could take off.

At 17:04:58, ATC reported another problem: the centre line lighting was out of service. Eleven seconds later, the voice recordings revealed that the Pan Am crew was still struggling to find the correct place to turn. By 17:05:44, Pan Am had actually missed its turning point and was still on runway 30.

Then occurred a crucial misunderstanding between ATC and KLM, perhaps exacerbated by the heavy Spanish accent of the Tenerife Air Controller falling on Dutch ears. Tenerife ATC cleared the KLM's take-off flight path, but did not clear the KLM to actually take off, because they weren't quite sure of the Pan Am's position.

Captain van Zanten now initiated take-off. After three and a half hours of delay, when he was on the verge of exceeding his legally allowable flight time, he was in a hurry. If he'd waited much longer, the KLM flight would have had to have been held over. Van Zanten's very seniority may have intimidated his much younger co-pilot, who failed to speak up. To make matters worse the communications system was overloaded with too many people trying to talk to each other at the same time. Between 17:06:19 and 17:06:23, communications caused a shrill noise in KLM cockpit and as a result the KLM crew didn't hear crucial messages.

The *Rhine River* started to take off. In the few seconds remaining, someone in the KLM cockpit suspected that something was going wrong (see box below), but it was too late.

At 17:06:40 the Pan Am captain saw the landing lights of the KLM Boeing about 700 metres away. Pan Am's First Officer Bragg: 'There he is ... Look at him! Goddam that son-of-a-bitch is coming! Get off! Get Off! Get Off!

FROM THE COCKPIT

17:06:32—KLM Flight Engineer: Is he not clear then?
17:06:34—KLM Captain: What are you saying?
17:06:34—KLM Flight Engineer: Is he not clear?
The Pan American?
17:06:35—KLM Captain: Yes. Definitely.

As the fog cleared from Tenerife the world discovered the full extent of the collision between two 747s.

WORDS OF A SURVIVOR

First Officer Robert Bragg set up a website on the Tenerife collision:

It's the only time in my life I have ever saw something happening that I could not believe was happening ... And I basically ducked, closed my eyes and when the KLM aeroplane hit us, I really didn't think the man had hurt us, it was a very slight impact, very slight noise, like CLUNK, that was about it, it was so minor it was unbelievable, until I opened my eyes and looked up. The first thing I noticed were that all of the windows were gone in the cockpit, then I looked out to the right and the right wing was on fire ... when I looked back the lounge and all of the people were gone.

I could see all the way to the tail of the aeroplane, just like someone had taken a big knife and sliced the entire top of the cabin off ... I looked up to grab the fire control handles. That's when I noticed that the entire top of the plane was gone, not only was the entire top of the plane gone ... there was only one foot [0.3 metres] of the floor left, the back of a pedestal, which is between the pilots, and we had two jump seats, directly back of the captain ... They were no longer there; those seats were gone. Turns out later that those gentlemen, they did the same thing as I did, they closed their eyes and ducked, when they looked up they were upside down, dangling from the ceiling of the aeroplane, dangling into the first class section, they were smart enough to reach over and grab the side of the aeroplane before they released their seat belts. They got out of the aeroplane and survived the accident.

The KLM crew immediately tried for a steep climb. The Pan Am crew applied full power for a left turn off the runway, but it was too late.

At 17:06:50, as the KLM 747 left the ground, its landing gear and engine clipped the Pan Am in the middle of the fuselage. The Dutch plane weighed 300 tons at this point and was travelling at 290 kilometres per hour (180 miles per hour). The impact broke the *Clipper Victor* into several large pieces. The *Rhine River* only climbed another 100 feet (30 metres) before van Zanten completely lost control. With full tanks the *Rhine River* crashed 250 metres (800 feet) down the runway and burst into flames, killing all onboard.

Seventy people survived the immediate collision on the *Clipper Victor*, but nine later died in hospital. Amazingly the Flight crew of Pan Am 1736 escaped the damaged plane, and lived to tell the tale.

In the aftermath of the Tenerife crash, authorities made important changes to the way that ground staff communicate with flight crew, including key requirements for a specific vocabulary to remove ambiguity. The industry now trains flight crew with the philosophy of crew resource management that emphasises group consensus for making decisions. Tenerife even went so far as to build another airport, Reina Sofia, in the southern part of the island, in order to avoid the fogs that frequent the north.

19 November 1977

Flight: TAP Air Portugal Flight 425
Model: Boeing 727-282
Registration: CS-TBR
Fatalities: 131/164 (125 out of the 156 passengers and 6 out of the 8 crew)
Principle cause: crew disorientation

Flight 425 had already abandoned two approaches to land at Funchal Airport, Madeira. Heavy rain made it impossible to see the runway. On the second attempt the flight crew thought they were pulling it off, but instead the 727 actually touched down 610 metres (2000 feet) past the runway threshold and had only 915 metres (3000 feet) of play left. It was far too little. At 21:35 the plane overshot the poorly drained runway and aquaplaned, until it plunged off a cliff and burst into flames.

1 January 1978

Flight: Air India Flight 855
Model: Boeing 747-237B
Registration: VT-EBD, *Emperor Ashoka*
Fatalities: 213/213 (190 passengers and all of the 23 crew)
Principle cause: instrument failure and pilot error

At 20:41, one minute after take-off from Mumbai, India, the *Emperor Ashoka* performed a gentle right turn, heading out towards the Arabian Sea on its way to Dubai. When the 747 levelled out, the pilot's horizon indicator still showed that the plane was banking right. In darkness and flying over the sea the pilot had no natural horizon to judge by and asked the co-pilot for an opinion. The co-pilot said that his horizon was fine, an opinion shared by the flight engineer who noticed the discrepancy. However, instead of actually listening to what his colleagues were telling him, the pilot continued to roll the plane left until the plane was at 40 degrees left. At this point the flight engineer spoke up, but the plane continued to roll until it reached 108 degrees, well on the way to being completely upside down. The crew lost control at 2000 feet (610 metres) and the 747 soon crashed into the Arabian Sea, off Bandra, India, in water only 10 metres (33 feet) deep.

'... instead of actually listening to what his colleagues were telling him, the pilot continued to roll the plane left ... until it reached 108 degrees, well on the way to being completely upside down.'

15 November 1978

Flight: Loftleidir Charter Flight
Model: McDonnell Douglas DC-8-63CF
Registration: TF-FLA
Fatalities: 183/262 (175 out of the 249 passengers and 8 out of the 13 crew)
Principle cause: flight crew error and possible instrument failure

On its return from Jeddah, Saudi Arabia, the DC-8 was making its final approach to Colombo, Sri Lanka, when it crashed into a coconut plantation at Katunavake, 1.6 kilometres (1 mile) short of the runway. The two competing versions of what caused the accident sound rather like a blame game.

The Sri Lankan report cited a number of failures on the part of the flight crew to conform to established landing procedures.

The Icelandic Directorate of Civil Aviation, who oversaw the operation of Loftleidir, cited the failure of the Sri Lankans to maintain their instrument landing system, which, they said, caused the Bandaranaike Airport radar controller to give the DC-8 crew fatally bad information.

25 May 1979

Flight: American Airlines Flight 191
Model: McDonnell Douglas DC-10-10
Registration: N110AA
Fatalities: 273/273 (260 passengers and 13 crew) and 2 on the ground, total 275
Principle cause: faulty design and poor maintenance

At 15:03 Flight 191 was fully loaded with fuel and about to begin a non-stop flight to Los Angeles. It had reached the rotation point (critical point of no return) of its take-off from Chicago-O'Hare, when sections of the number one engine pylon began to fall off. During rotation engine one separated from the wing and rolled over it before falling on to the runway.

Air traffic control frantically radioed Captain Walter Lux to ask him if he wanted to abort the take-off, but it was already too late. The engine loss had caused an electrical failure, which destroyed the cockpit controls, and there was nothing the flight crew could do.

At the 1800 metre (6000 foot) mark, the plane lifted off and momentarily reached a level flight at 300 feet (90 metres), before it started rolling to the left. The nose pitched down and the left roll continued until the wings were at vertical. At 15:04 the DC-10 crashed into a trailer park, 1430 metres (4680 feet) northwest of the end of the runway. As the plane exploded it killed two people on the ground and two others sustained major burns.

Blinding rain, poor visibility and inadequate runway drainage proved a fatal combination for Air Portugal Flight 425.

The wreckage of a Swissair DC-8 which skidded at the end of the runway at Athens airport on 11 October 1979 and burst into flames, killing fourteen passengers and hospitalising thirty others.

The National Transport Safety Board (NTSB) report was scathing, but fair. It cited: the failure of McDonnell Douglas to design engine support pylons resistant to 'maintenance-induced damage'; the failure of the Federal Aviation Administration's surveillance and reporting systems; and the failure of American Airlines to take better care of its planes.

At the time DC-10s were fitted with closed-circuit televisions, which allowed the passengers to see the flight from the cockpit's point-of-view. The possibility that the passengers were watching their own deaths on television added a final, surreal note to the whole debacle.

> 'The possibility that the passengers were watching their own deaths on television added a final, surreal note to the whole debacle.'

This was the worst air disaster in American history until the 9/11 World Trade Centre attacks.

28 November 1979

Flight: Air New Zealand Flight 901
Model: McDonnell Douglas DC-10-30
Registration: ZK-NZP
Fatalities: 257/257 (237 passengers and 20 crew)
Principle cause: navigational or pilot error

This crash is notable for the fact that Sir Edmund Hillary—the first man, along with Tenzing Norgay, to scale Mount Everest—had originally been scheduled to act as a guide on the trip, as he had on other flights, but had to cancel because of other commitments. His friend Peter Mulgrew was an experienced explorer in his own right and replaced him.

In the days before comprehensive satellite navigation systems, aeroplanes had to rely partly on magnetic compasses to know which way was north or south. However, Air New Zealand's popular sightseeing trips to Antarctica took the planes too close to the Southern Magnetic Pole to maintain reliability so aircrew had to rely on another navigational method.

The inertial navigational system (INS) relied on brute force arithmetic. It simply calculated the position of the plane based on the sum of all speeds and durations it had flown and the directions that it had flown in. The INS could in fact be programmed in advance to take the plane through a prescribed route. The only flaw in the system was that whenever you changed a program that people were used to, you had to tell the people involved so that they'd know where they were.

Nineteen days before the flight, Air NZ staff had briefed Captain Jim Collins and Co-Pilot Greg Cassin about their route. At 08:21 on the day of the flight they programmed the INS. Unfortunately, no one had told them that someone had altered the flight path slightly, so the data that the pilots were programming in didn't match what they were imagining in their heads.

Flight 901 took off from Auckland at 07:17 on 28 November. By 12:18 the crew thought that they had reached McMurdo Sound. Unfortunately, low-lying clouds were obscuring the view that the passengers had paid so much and had travelled so far to see. Even worse was the fact that they weren't over McMurdo Sound at all but 2.4 kilometres (1.5 miles) east over Lewis Bay. Lewis Bay looks a little like McMurdo, but out of all of the crew, only one of the flight engineers had been there. Everyone else was relying on the pre-programmed INS. The crew decided to fly lower to have a better look—lower, in fact, than their authorised height of 16 000 feet (4900 metres).

Half an hour later, the crew was really confused and the black box records them discussing the matter with Peter Mulgrew. Crucially, they couldn't work out where Mount Erebus was in relation to the plane. Deciding that things didn't 'look very good at all', they decided to turn right and climb out of the area.

Their instruments told them that they were hundreds of metres over flat ground, but as they made their turn all hell broke loose and the ground proximity warning system (GPWS) immediately sounded its warning of 'WHOOP, WHOOP! PULL UP! WHOOP WHOOP!' The crew must have thought that it was a mistake, but their training taught them to listen to the machines. However, the GPWS wasn't designed to operate near the sheer faces of mountains and the crew barely had any time to react.

The captain decided to deploy go-around power, which applied the full thrust and lift that was usually only ever needed at take-off—a desperate measure, but it was too late. At 12:50 Flight 901 slammed into the side of Mount Erebus and disintegrated.

The following year the official report of the New Zealand Chief Inspector of Aircraft Accidents attributed the crash to:

The decision of the captain to continue the flight at low level toward an area of poor surface and

A TRAGIC MILESTONE

In a country of only three million people and with the loss of so many lives, practically everyone in New Zealand was in some way personally involved.

horizon definition when the crew was not certain of their position and the subsequent inability to detect the rising terrain which intercepted the aircraft's flight path.

However, this conclusion wasn't good enough for the general public, and they demanded another enquiry. In 1981 Justice Peter Mahon gave his version of what happened in his one-man Royal Commission into the crash: 'The dominant cause of the disaster was the act of the airline in changing the computer track of the aircraft without telling the air crew …' Justice Mahon went further and publicly accused Air New Zealand of a cover-up, calling their testimony to the earlier report 'an orchestrated litany of lies'.

Air New Zealand had to pay for the costs of the Mahon Commission and an extra fee of NZ$150 000. The CEO of Air New Zealand resigned within a week. The case of Flight 901 remains highly unusual, in that conclusions about who was responsible and who was to blame polarised public opinion. Peter Mahon is something of a folk hero in some quarters.

An aluminium cross now stands at Scott Base as a memorial. Almost all of the wreckage of the DC-10 still clings to the sides of Mount Erebus. In the winter, snow and ice bury it and hide it, but later in the year, the snow melts and from high above you can see the shards of jagged metal shining in the Antarctic summer sun.

A military plane lands in Antarctica with Mount Erebus in the background. It's difficult to imagine, on a clear day, how a flight crew can fail to see a mountain. All the more reason to understand how disorienting low-lying cloud can be—and how fatal.

Tehran's Mehrabad airport, the site of another horrific accident in 2000, between two Iranian planes, a Hercules C-130 transport and an IranAir Airbus A-300.

21 January 1980

Flight: Iran National Airlines
Model: Boeing 727-86
Registration: EP-IRD, *Shiraz*
Fatalities: 146/146 (138 passengers and 8 crew)
Principle cause: navigational equipment failure

The *Shiraz* was approaching Tehran, Iran, at night and in fog and snow, when the airport's instrument landing system and ground radar were inoperable. The crew had to rely on visual flight rules in conditions of compromised visibility and became disoriented. At 19:11 the *Shiraz* crashed in the Elburz Mountains, north of Tehran.

25 April 1980

Flight: Dan Air Flight 1008
Model: Boeing 727-64
Registration: G-BDAN
Fatalities: 146/146 (138 passengers and 8 crew)
Principle cause: navigational error

Flight 1008 had originated in Manchester and was due to land at Los Rodeos Airport when air traffic control instructed it to hold at 5000 feet (1524 metres) around the Airport's FP beacon. Instead the pilots took the plane 5 kilometres (3 miles) south of the beacon.

The official reports stated that, 'The captain, without taking into account the altitude at which he was flying, took the aircraft into an area of very high ground, and for this reason he did not maintain the correct safety distance above the ground, as was his obligation.' As a result, the 727 crashed into a mountain at a height of 1661 metres (5450 feet).

Dan Air no longer exists. British Airways bought it on 23 November 1992 for the princely sum of £1—that's $3.90 in today's terms, even using the most generous currency conversions.

1 December 1981

Flight: Inex Adria Avioproment
Model: McDonnell Douglas DC-9-81
Registration: YU-ANA
Fatalities: 180/180 (173 passengers and 7 crew)
Principle cause: miscommunication

The DC-9 was on approach to Ajaccio-Campo Dell'Oro Airport on Corsica and almost at the end of its flight from Ljubljana-Brnik Airport in Slovenia. The plane was in a holding pattern, but air traffic control thought it was going for a direct descent and gave instructions accordingly. The flight crew complied with ATC and descended below the minimum holding pattern of 6800 feet (2075 metres), only to have their ground proximity warning system go off. Although they had thirteen seconds to react, the pilots responded too late and the plane collided near the top of 1364 metre (4477 foot) high Mount San-Pietro, south west of Ajaccio.

13 January 1982

Flight: Air Florida Flight 90
Model: Boeing 737-222
Registration: N62AF
Fatalities: 74/79 (70 out of the 74 passengers and 4 out of the 5 crew) and 4 on the ground, total 83
Principle cause: poor weather conditions and pilot error

Early 1982 gave the entire east coast of America record low temperatures, so it was a cold winter's day in Washington. The airport had closed shortly after the 737 had arrived, but as it was originally scheduled to reopen at 14:30, Flight 90's captain Larry Wheaton ordered de-icing. Later investigation revealed that maintenance staff did not cover any airframe openings during the procedure, allowing ice to form in dangerous places.

When the airport reopened at 14:53, the combination of ice, snow and glycol on the tarmac had left the tug without enough traction to move the plane. Wheaton decided to use reverse thrust to shift the 737—a move contrary to flight manual procedures. This only resulted in blowing snow into the air, some of which may have ended up on the fuselage. The tug tried again and finally succeeded in pulling the plane onto the runway by 15:38. Flight 95 finally seemed to be closer to its intended destination, Fort Lauderdale, Florida.

Take-offs from Washington are difficult at the best of times. The initial flight path requires tricky negotiation around the restricted airspace over the Washington Monument and The Pentagon and on that day and in blizzard conditions there was only one instrument-rated runway. The planes had to queue to take off, and all the while the snow continued to fall, rapidly undoing whatever good the de-icing had done. To make matters worse, the plane's own anti-icing systems were not on. The crew was curiously blasé about this especially since, based in Florida, they had very little experience with freezing conditions.

In another unorthodox move the flight crew attempted to de-ice their plane by using the exhaust of the New York Air DC-9 in front of it. Investigators later speculated that all this may have done was to melt the snow onto the wing's leading edges, which could have blocked the engines—even a little ice was known to seriously affect the performance of 737s.

At 15:58 Flight 90 received take-off clearance. Something must have alerted the first officer, because, as the 737 initiated take-off, the co-pilot repeated four times to the captain that something wasn't right, yet Larry Wheaton made no attempt to abort.

The plane took longer and further to accelerate than normal, failed to accelerate after take-off, then stalled. At 16:01, after having attained a maximum altitude of just 337 feet (103 metres), the 737 practically nosedived into the busy and congested 14th Street Bridge, which spanned the Potomac River just north of the airport.

The plane struck six moving cars and a boom truck, before plunging into the freezing river. Autopsies later revealed that impact injuries killed the passengers; none drowned immediately. Some of the most dramatic rescue footage ever taken recorded what happened next.

Flight attendant Kelly Duncan and five passengers, the only survivors, were clinging to the tail section of the downed plane. Kelly gave the only life vest they could find to passenger Nikki Felch.

In the massive back-up of traffic from the blizzard, it was almost impossible for ambulances to reach the crash site, so it was down to civilian bystanders to effect a rescue. Without adequate equipment they made a rope from scarves, battery cables and anything else they could find. Then a sheet metal foreman, Roger Olian, attached the rope to himself and dived into the river to try to reach the survivors. He spent twenty minutes in ice water doing his best—an amazing feat, since people can die from hypothermia in such conditions in only a few minutes.

Then the only rescue helicopter arrived. At 16:20 the helicopter lowered a line and managed to pull survivors to shore—Bert Hamilton first, then Kelly Duncan, then a severely injured Joe Stiley. Two female passengers Patricia Felch and Priscilla Tiraldo then took the line. Both women lost their grip and fell into the river. Another civilian, Lenny Skutnik, leapt into the water to save them, as did Gene Windsor out of the helicopter, in order to attach the line to Priscilla.

All this time another passenger, 46-year-old bank examiner Arland Williams, had graciously handed the lifelines to his more seriously injured companions. He had been clinging to the tail for almost half an hour. By the time the helicopter lifeline was ready for him the tail had sunk into the Potomac. Arland Williams was the only passenger to die from drowning. At the time, no one had yet identified him, and the following day *The Washington Post* published a story about the 'sixth passenger', entitled 'A Hero—Passenger Aids Others, Then Dies'.

Olian, Skutnik, Usher and Windsor all received awards and medals in recognition of the work they did that day, but perhaps the most poignant memorial went to Williams.

Very cold weather creates such a demanding environment that planes require special fuel mixes to operate in low temperature environments.

Planes may be relatively fragile, but they can still take a beating. Survivors tell us that near misses are, nevertheless, absolutely terrifying.

After its repair 14th Street Bridge was officially renamed the Arland D Williams Jr Memorial Bridge in his honour.

8 June 1982

Flight: VASP Flight 168
Model: Boeing 727-212
Registration: PP-SRK
Fatalities: 137/137 (128 passengers and 9 crew)
Principle cause: pilot error

Flight 168 had originated in Sao Paulo and was on its final night-time approach to Fortaleza on the north coast of Brazil. Air traffic control cleared the crew to descend to 5000 feet (1525 metres). Instead of stopping at 5000 feet, the pilot continued his descent, even ignoring two ground proximity warning system alerts as well as his co-pilot's warning of mountains ahead. At 02:25 the 727 crashed into elevated woodlands in the Sierra de Pacatuba killing all onboard.

'... the pilot continued his descent, even ignoring two ground proximity warning system alerts as well as his co-pilot's warning of mountains ahead.'

28 June 1982

Flight: Aeroflot Flight 8641
Model: Yakovlev 42
Registration: CCCP-42529
Fatalities: 132/132 (124 passengers and 8 crew)
Principle cause: systems failure

Less than three weeks after the VASP disaster, another accident hit the current list of the top 101 worst accidents. Metal fatigue in the jackscrew of the horizontal stabiliser of Flight 8641 caused the component to fail while the plane was en route from Leningrad to Kiev, Russia. The crew lost control of the plane and at 10:50 crashed just south of the city of Mozyr in Belarus, killing all onboard.

In the aftermath of the tragedy, authorities grounded all Yakolev 42s, and the planes underwent major design changes until deemed worthy to fly again in late 1984.

27 November 1983

Flight: Avianca Flight 11
Model: Boeing 747-283B
Registration: HK-2910
Fatalities: 181/192 (162 out of the 173 passengers and all of the 19 crew)
Principle cause: navigational error

Shortly after midnight, as Flight 11 was making its final approach to Madrid, the flight crew entered the airport's instrument landing system incorrectly and keyed in the wrong coordinates for the airport's radio-range tower into the plane's navigation system. As a result the 747 made a turn short of the tower. The right main landing gear and number four engine hit one hillside, then the plane hit a second hill. Six seconds after that the plane was still travelling at 233 kilometres per hour (145 miles per hour) when it hit the ground with its right wing, which tore off. At 00:06 the jumbo cartwheeled and broke into five pieces before the fuselage came to rest upside down 12 kilometres (7.5 miles) south east of Madrid-Barajas airport.

FAMOUS CASUALTIES

Onboard Avianca Flight 11 on 27 November 1983 had been several key South American cultural figures, including Mexican novelist Jorge Ibarguengoitia, Uruguayan novelist Angel Rama, Argentinian art critic Marta Traba and Peruvian poet Manuel Scorza.

10 July 1985

Flight: Aeroflot Flight 7425
Model: Tupolev 154B-2
Registration: CCCP-85311
Fatalities: 200/200 (191 passengers and 9 crew)
Principle cause: pilot error

The higher up in the atmosphere you go, the thinner the air becomes. Because of this aeroplanes have a maximum operational ceiling—beyond a certain height the plane simply doesn't have enough air going over the wing to maintain the vacuum that keeps the plane aloft. How high this ceiling is depends to some extent on the plane's weight. The lighter the aircraft, the higher it can go.

The Tupolev was en route from Karshi, Uzbekistan, to Orenburg, Russia, when it encountered a thunderstorm near Uch-Kuduk. The pilot wanted to climb to 38 000 feet (11 580 metres) to avoid the rain and turbulence. Unfortunately, in doing so he forgot the basic laws of physics and the jet climbed too high for its weight. The airspeed dropped to 290 kilometres per hour (180 miles per hour), the plane stalled, went into a flat spin and crashed, killing all onboard.

12 August 1985

Flight: Japan Air Lines Flight 123
Model: Boeing 747-SR46
Registration: JA8119
Fatalities: 520/524 (509 passengers and 15 crew)
Principle cause: structural collapse

On 2 June 1978, a full seven years and ten weeks before its last flight, the plane had touched down at Osaka and floated up again for a short time. On its second landing, the tail had struck the runway, damaging the fuselage and the rear pressure bulkhead. During repairs JAL and Boeing replaced the damaged sections. However, the investigation of the plane's final flight later revealed that the repair work was shoddy and had been done with only half the necessary rivets. The plane flew for years and the only indication that anything was wrong was that the tail would make occasional whistling noises, but no one at JAL ever did anything about it.

Hundreds of flights later, on 12 August 1985, the 747 left Tokyo, bound for Osaka. At 18:24 the plane reached an altitude of 23 900 feet (7285 metres) and was travelling at 555 kilometres per hour (345 miles per hour) over Sagami Bay when it started to shudder violently and underwent an explosive decompression. The crew thought that the problem was with a door at the rear. What had actually happened was that metal fatigue had caused the rear bulkhead to rupture, a 5-metre (16 foot) section of the vertical tail fin to come loose and the section of the tail cone that contained the auxiliary power unit to break off. Hydraulic pressure dropped; ailerons, elevators and yaw dampers failed and the plane became virtually unflyable. The plane descended and flew over the Izu Peninsula, heading out towards the Pacific Ocean then back to land. It wandered aimlessly and completely out of control for over half an hour. The black box transcript reveals little but the complete helplessness of the pilots.

All the time that the plane was drifting, the passengers knew that they were doomed. In their final moments, many composed parting letters and poems to their loved ones. Rescuers found these poignant last messages later at the crash site written in shaky handwriting, struggling to remain steady as the plane continued its wild oscillations:

A TRAGIC MILESTONE

The SR series of 747s is specifically designed for 'short runs' with high volume traffic, perfect for the domestic Japanese market, but it is this fact which explains why this accident was the worst in aviation history involving a single plane—a record that it still holds over twenty years later. The jumbo was full to capacity and about as full as a 747 could be at the time.

My Darling Wife,
Life with you has been wonderful. May our children grow up to be people that I can be proud of. I never dreamed that the dinner we had last night would be our last one together.

The plane descended to below 7000 feet (2100 metres) when the pilots gained a faint glimmer of hope, but after the plane reached an altitude of 13 000 feet (4000 metres) it began a steep, almost vertical and unrecoverable descent.

At 18:56, 45 minutes into its flight and 100 kilometres (62 miles) from Tokyo, JAL 123 struck a tree-covered ridge near Ueno. Three seconds later it crashed into the forest of Mount Otusuka.

Twenty minutes after the crash an American Air Force helicopter visited the site and radioed colleagues at Yokota Air Base to mobilise and help with the rescue, but Japanese authorities ordered the Americans back to their base and sent a team from the Japan Self-Defence Force. By then visibility in the mountains was zero and rescue teams had to wait until morning to do anything further. Some passengers may have survived the crash, but succumbed to hypothermia in the cold mountain air. One off-duty stewardess, Yumi Ochiai, survived and later reported that, when she first roused from the impact, she could hear the screaming and moaning of other survivors, which gradually died down during the night.

The poignant remains of the worst air disaster ever involving a single aircraft—so far. 520 lives ended on a Japanese mountainside.

A soldier salutes, honouring the dead. In late 1985, 248 members of the US 101st Airborne Division died when their Douglas DC-8 went down in the Sinai Desert. The cause remains a mystery.

Three other passengers also made it through —a twelve-year-old girl who rescuers found in a tree, a mother and her eight-year-old daughter. All the women had been sitting in the tail section of the plane and were suffering from broken bones and other injuries.

After the crash several of those responsible committed suicide, including both the Boeing engineer who had performed the faulty repair work and also the JAL official who had approved the original repair work seven years before. Boeing and JAL settled on an 80/20- liability split respectively and settlements for the families came out of insurance money.

'After the crash several of those responsible committed suicide ...'

12 December 1985

Flight: Arrow Airways Flight MF 1285R
Model: Douglas DC-8-63CF
Registration: N950JW
Fatalities: 256/256 (248 passengers and 8 crew)
Principle cause: 'not definitively determined'

Multinational Force and Observers (MFO) had chartered the DC-8 to take 248 members of the 101st Airborne Division and their equipment from a peacekeeping mission in the Sinai Desert. The flight departed from Cologne Airport at 03:50 bound for Gander, Newfoundland, and arrived there at 05:34 without incident. At 06:45 the plane began its take-off and reached an airspeed of 309 kilometres per hour (192 miles per hour) when it began rotation. It took off, barely, and crossed the Trans-Canada Highway at low altitude. Although the plane's pitch angle increased, it nonetheless continued to descend. At 06:46 it struck the ground, broke up and burst into flames. The plane had made it only 915 metres (3000 feet) beyond the end of the runway.

In its official report, the Canadian Aviation Safety Board (CASB):

... was unable to determine the exact sequence of events which led to this accident. The Board believes, however, that the weight of evidence supports the conclusion that, shortly after lift- off ... the most probable cause of the stall was determined to be ice contamination on the leading edge and upper surface of the wing.

However, four CASB members filed a dissenting opinion, citing, 'An in-flight fire that may have resulted from detonations of undetermined origin (that) brought about catastrophic system failures.'

31 March 1986

Flight: Mexicana Flight 940
Model: Boeing 727-264
Registration: XA-MEM
Fatalities: 167/167 (159 passengers and 8 crew)
Principle cause: maintenance error

At 08:40, during take-off from Mexico City to Vallarta, Mexico, Flight 940's left main gear brake overheated. The gear retracted normally, but when the plane had reached an altitude of 31 000 feet (9450 metres), a tyre on the gear exploded. The explosion severed electrical cables and ruptured fuel and hydraulic lines as well as causing cabin decompression. The crew declared an emergency, but leaking fuel ignited and flames engulfed the plane. The crew lost control and the 727 crashed into a mountain in Las Mesas, in the Sierra Madre.

Subsequent investigation revealed that a maintenance worker had incorrectly filled the tyre with air instead of nitrogen. Under high temperature and pressure, the oxygen in the air reacted chemically with the rubber in the tyre and the explosion killed everyone onboard Flight 940.

17 March 1988

Flight: Avianca Flight 410
Model: Boeing 727-21
Registration: HK-1716
Fatalities: 143/143 (136 passengers and 7 crew)
Principle cause: pilot error

Human relations problems in the cockpit seem to be what brought down this plane. One of the pilots had invited a non-crew member onto the flight deck and the conversation seemed to have affected discipline and teamwork. The plane was also late on its run to Cartagena, Colombia, and the crew may have been suffering from 'get-there-itis'—the peculiar condition in which the desire to make schedule compromises proper procedure and safety. The crew had also taken off from Cucuta using visual flight rules in foggy instrument flight rules conditions. When combined with the disciplinary problems, Flight 410 was asking for trouble; it found it when the plane struck a mountain north of Santander, Columbia, at 13:17.

' ... the crew may have been suffering from 'get-there-itis'—the peculiar condition in which the desire to make schedule compromises proper procedure and safety.'

2 February 1989

Flight: Independent Air Inc Flight 1851
Model: Boeing 707-331B
Registration: N7231T
Fatalities: 144/144 (137 passengers and 7 crew)
Principle cause: pilot error/miscommunication

Flight 1851 had flown from Bergamo, Italy, and was on its final approach to Santa Maria in the Azores. Santa Maria ATC was clearing the 707 for a descent to 3000 feet (915 metres):

Independent Air one eight five one roger, reclear to three thousand feet on QNH one zero two seven and runway will be one niner … [pause] … expect ILS approach runway one niner report reaching three thousand.

In the brief pause between 'niner' and 'expect' the co-pilot interrupted with 'Reclear to two thousand feet and ah … [pause] … one zero two seven.'

The Board of Inquiry investigation cited that what happened next:

... was due to the non-observance by the crew of established operating procedures, which led to the deliberate descent of the aircraft to 2000 feet in violation the minimum sector altitude of 3000 feet, published in the appropriate aeronautical charts and cleared by the Santa Maria Aerodrome Control Tower.

Due to the co-pilot's premature response, air traffic control didn't hear the pilots' mistake, and the pilots didn't hear the correct approach. The plane was now flying 1000 feet (300 metres) lower than it ought to have been. The ground warning proximity system sounded without comment or reaction from the crew and, at 14:08, Flight 1851 crashed into Mount Pico Alto.

Avianca, the Colombian airline, has been the reluctant host to some of the most dramatic crashes in civil aviation history.

A United Airlines plane takes off from Denver. Another of the world's 90 000 daily commercial flights that don't end in disaster.

7 June 1989

Flight: Surinam Airways Flight 764
Model: McDonnell Douglas DC-8-62
Registration: N1809E
Fatalities: 176/187 (167 of 178 passengers and all 9 crew)
Principle cause: pilot error

Flight 764 was on its final approach to Paramaribo, Suriname, at the end of its flight from Amsterdam. ATC informed the pilots that there was low fog, but had to clear them for visual approach because the instrument landing system wasn't working properly. The crew ignored this and went into an instrument landing anyway.

The investigating commission later determined 'That as a result of the captain's glaring carelessness and recklessness the aircraft was flown below the published minimum altitudes during the approach ...' The ILS signal was unreliable, the plane came in too low and its number two engine clipped at tree at a height of 82 feet (25 metres), the right wing then struck another tree and the DC-8 rolled and crashed upside down.

19 July 1989

Flight: United Airlines Flight 232
Model: McDonnell Douglas DC-10-10
Registration: N1819U
Fatalities: 111/296 (110 out of the 285 passengers and 1 out of the 11 crew)
Principle cause: mechanical failure

At 15:15 the DC-10 was on its way from Denver to Chicago and cruising at 37 000 feet (11 300 metres) over an Iowa cornfield, north-east of Sioux City, when the plane's number two, tail-mounted, engine suffered an uncontained failure. The investigation later revealed that a flaw the size of the a grain of sand had been in the engine's fan disc ever since General Electric had manufactured it. Ultrasonic examination had failed to pick it up and the NTSB also later criticised United Airlines' maintenance for failing to find the 'hard alpha inclusion stress fracture' that developed over time.

One hour and seven minutes into Flight 232's last trip, the fracture broke the fan disc in two, spun the fragments out and sent high-speed shrapnel into the tail assembly. The damage broke off the engine casing and tail cone and severed vital control hydraulic lines, making the plane virtually unflyable.

The Sorenson family later recovered the disc in their cornfield and they donated GE's $120 000 reward money to charity ($200 000 in today's terms).

Dennis Fitch was a United DC-10 flight instructor deadheading (catching a free lift) on the plane at the time. He turned out to be exactly the right person in the right place at the wrong time. True to the warning that those who ignore history are destined to repeat its mistakes, Denny Fitch had extensively studied the Japan Air Lines Flight 123 crash almost four years before.

When he went to the flight deck to help the flight crew, he was able to give them valuable advice that saved scores of lives. The plane had begun an uncontrollable tendency to turn right, both sets of wing elevators (the ailerons) were now up. The plane had begun weaving up and down every minute and was losing altitude all the while. Denny helped the pilots to achieve some stability and rough steering using the thrust of the remaining engines. The flight path for the next 45 minutes was a triumph of skill and luck over gravity, as the flight crew struggled to reach the nearest airport, Sioux Gateway in Iowa.

At 16:00, however, most of their luck ran out. Just as the plane reached the runway, it was falling at a rate of 1800 feet (550 metres) per minute and hit the runway.

WORDS OF A SURVIVOR

Senior flight attendant Janet Brown Lohr described those awful seconds:

It was horrendous these strange, shrieking noises ... I couldn't hear anything. I couldn't feel anything. I couldn't smell anything. Nothing was working except my mind. It was like total body detachment or being in a protective cocoon. Then I realised that two-thirds of me was suspended in fire and I felt, 'This is it. This is how I'm going to go. This is how I'm going to die', and it was the most incredibly peaceful moment I've ever known.

I opened my eyes and I couldn't recognise a thing. It was like waking up on a foreign planet and I'm thinking, 'I'm still thinking. In other words, I'm still alive.' Then it was right back to work. We've got to get out of here.

Footage taken at the time of the impact shows a plane tumbling to pieces and turning into a fireball.

At least some of the luck had held. The fact that almost 200 people survived this destruction gave new meaning to the term 'miracle', and although the airport had been forewarned, the rescue was still far from easy.

Denny later said that he 'cried for three days, on and off' and credited the amazingly high survivability of what should have been an unsurvivable crash to three crucial facts:

1) The crash had occurred in daylight.
2) The crash had occurred during a shift change at Sioux City's regional burn and trauma centres, virtually doubling the available staff at shortnotice.
3) The Iowa Air National Guard was on duty at Sioux City Airport at the time, making an extra 285 triage-trained professionals available.

Others were equally happy to credit Denny Fitch. He recovered from the extensive injuries he received during the crash and is now a safety consultant with NASA, as well as running his own consulting firm. He was inducted into the National Aviation Hall of Fame at the Smoothening National Air and Space Museum and continues to give motivational speeches based on his experiences.

25 January 1990

Flight: Avianca Flight 052
Model: Boeing 707-321B
Registration: HK-2016
Fatalities: 73/158 (65 out of the 149 passengers and 8 out of the 9 crew)
Principle cause: miscommunication

By any reasonable standards this is an accident that should not have happened. Flight 052 had left Medellin Airport in Colombia at 15:08 with enough fuel to take it to New York and still keep it in the air for two more hours. The 707's captain Laureano Caviades had only poor English and relied on his junior co-pilot Mauricio Closs to communicate. Closs may have found Caviades' experience intimidating, which may have contributed to what eventually happened.

At 20:00 a massive low-pressure system had built up along the entire east coast of America, travelling west to east. Washington DC air traffic managers were aware of the deteriorating weather, but still ordered JFK Airport, New York, to land 33 planes per hour, even though it was near blackout and had only one operating runway.

Various air traffic controls then put Flight 052 into a series of holding patterns: over Norfolk, North Virginia, for nineteen minutes; over New Jersey for 29 minutes; and over JFK Airport for 26 minutes. At 20:44 ATC New York advised Flight 052 that they would receive further clearance at 21:05. Only now did the crew

Poor communication on Flight 052 did nothing to improve Avianca's safety record.

'... even though the captain ordered the co-pilot to advise ATC
that they were in an emergency, the co-pilot didn't actually do it ...'

Residents of nearby villages look at the wreckage of the Lauda-Air Flight 004 which crashed on 26 May 1991 in Thailand.

ATC that they could only really hold for another five minutes—an alternate landing at Boston was now out of reach. Flight 052 left the holding pattern three minutes later for their final approach, but they had no room to manoeuvre.

When they were finally cleared for a landing at 21:19, gusting headwinds and high-altitude windshear slowed them down and forced them to use full throttle—and a lot of fuel. They had to abort a landing at 21:23. Visibility in the thunderstorm was so low that the captain even commented: 'I don't know what happened to the runway, I didn't see it.'

Then, even though the captain ordered the co-pilot to advise ATC that they were in an emergency, the co-pilot didn't actually do it, even though he told the captain that he had. At no time did he use the term 'emergency'. In translating from the Spanish *prioridad* to the English, he used the phrase 'we need priority'. It seems that, in the combination of stress and fatigue, the pilots simply failed to communicate clearly either to each other or to ATC. This is in spite of the fact that the Spanish word *emergencia* is almost identical in form and meaning to 'emergency'.

On the next turn around the plane ran out of fuel. What little was left in the tank couldn't reach the feeds, because when the plane's nose went up the fuel sloshed to the back of the wing tank.

At 21:32, 19 kilometres (12 miles) south-east of JFK Airport, engines four then three flamed out, then one and two. The cabin lights went out and the plane went down two minutes later on a hillside on the north shore of Long Island. A local resident called 911: 'I live in Cove Neck in Oyster Bay. There's a plane in our yard, in front of our house.' The plane had just missed destroying her house and had fractured the wooden deck.

The one positive aspect of the tragedy that killed 73 people, including the flight crew, was that in the absence of fuel there was no fire. As it was, just over half the people on the 707 escaped the wreckage alive.

26 May 1991

Flight: Lauda Air Flight 004
Model: Boeing 767
Registration: *Mozart* OE-LAN
Fatalities: 223/223 (213 passengers and 10 crew)
Principle cause: systems failure

The *Mozart* took off from Bangkok Airport at 23:10 bound for Wien-Schwechat Airport in Vienna. At 23:22 the plane was climbing, when the plane's systems advised them that a system failure might activate the thrust reverser of the port engine. The thrust reverser is the plane's 'reverse' gear. When a plane touches down, it has to lose speed very quickly or it will run out of runway. A thrust reversal in-flight is a disaster. One engine would be suddenly thrusting backwards, while all of the others would still be thrusting forwards.

FROM THE COCKPIT

Co-pilot: You need a little bit of rudder trim to the left.
Captain: Ah, reverser's deployed.
(Ambient sound in cockpit of a snap)
Captain: Jesus Christ!
(Cockpit sound of four caution tones. Siren warning starts, stops and starts again)
Captain: Here! Wait a minute! Damn it!
(Sound of a bang)

The forces involved would not only make the plane uncontrollable, it would break the plane apart in seconds.

The crew thought that the signal itself was a malfunction. Boeing's *Emergency Malfunction Checklist* even told them 'No Action Required'.

At 23:31, when the *Mozart* was at 4000 feet (1220 metres) engine one's reverse thruster isolation valve failed and the thrusters deployed. No amount of piloting skill could have recovered from what happened next. The plane and its passengers didn't stand a chance.

The fuselage crashed into the ground at Ban Non Waeng in Thailand. The cockpit and one wing landed in separate places, each over a mile from the main crash site.

28 September 1992

Flight: Pakistan International Airlines (PIA) Flight 268
Model: Airbus A300 B4-203
Registration: AP-BCP
Fatalities: 167/167 (155 passengers and 12 crew)
Principle cause: pilot error

Kathmandu Airport is situated in a long, oval-shaped valley at an altitude of 4390 feet (1343 metres), surrounded by mountains as high as high as 9665 feet (2346 metres). The approach is difficult and steep, and it requires flight crews to reach certain critical altitudes at critical points. Flight 268 was on its final approach when the crew mistook one altitude nexus for another, and at 14:30 the Airbus A300 crashed into a steep, cloud-enshrouded mountain at an altitude of 7300 feet (2225 metres), 20 kilometres (12.5 miles) south of the Nepalese capital.

24 November 1992

Flight: China Southern Airlines Flight 3943
Model: Boeing 737-3YO
Registration: B2523
Fatalities: 141/141 (131 passengers and 10 crew)
Principle cause: mechanical failure and pilot error

Flight 3943 had originated in Guangzhou, China, and at 07:50 was about to land at Guilin. The crew had the auto-throttle engaged. As the captain attempted to level off the plane, the auto-throttle failed to activate engine number two. The manoeuvre required that the plane reach a particular airspeed. The automatic systems now registered that there was not enough power and compensated by increasing the power to engine number one. Thrust was now asymmetric and the plane veered right when it was low to the ground. At 07:52 the 737 crashed near Liutang, 20 kilometres (12.5 miles) south of Guilin.

19 September 1993

Flight: SAM Colombia 501
Model: Boeing 727-46
Registration: HK-2422X
Fatalities: 132/132 (125 passengers and 7 crew)
Principle cause: sabotage and instrument failure

Terrorists had attacked the radar system at Medellin, Colombia, so that when the flight from Panama was making its final approach in a thunderstorm, the 727 was flying in too low with compromised visibility. At 15:04 the plane crashed into Mount Paramo Frontino.

26 April 1994

Flight: China Airlines Flight 140
Model: Airbus A300B4-622R
Registration: B-1816
Fatalities: 264/271 (249 of 256 passengers and 15 crew)
Principle cause: pilot error

In perfect weather conditions the Airbus was approaching Nagoya, Japan, when the co-pilot inadvertently selected Take Off/Go Around Mode (TOGA). The rest of the crew failed to pick this up, and, when the plane refused to configure itself for landing, the crew tried to override the automatic controls.

However, as the plane came close to landing, the pilots re-engaged the automatics, which were still preset to TOGA, so instead of adjusting for landing, the plane was trying to take off again, even as the crew's inputs were telling it to touch down. The plane's systems were now receiving contradictory inputs. As sensors noted that the plane's pitch and angle of attack were too high, the systems compensated by increasing thrust only to have the nose-up attitude rise to 52.6 degrees.

After authorities have cleared a crash site of possible dangers, the question that occurs to recovery teams is: 'where to begin?' The remains of the China Airlines Flight 140 Airbus that crashed 26 April in Nagoya, Japan, killing 265 people with seven survivors.

A desperate firefighter uses a tree branch to beat out a fire after a Boeing 707 cargo jet crashed into homes near the La Aurora International Airport, Guatemala.

By now the airspeed was down to just 145 kilometres per hour (90 miles per hour) and the plane stalled. At 20:16 it hit the ground tail-first and collapsed into a fireball.

6 June 1994

Flight: China Northwest Airlines Flight 2303
Model: Tupolev 154M
Registration: B-2610
Fatalities: 160/160 (146 passengers and 14 crew)
Principle cause: maintenance error

The evening before Flight 2303's planned trip from Xian, China, to Guangzhou the B-2610 had undergone some maintenance in the field. Ground staff had inadvertently connected the autopilot's yaw channel to the roll channel and vice-versa. As a result, at 08:20, shortly after take-off when the crew engaged the autopilot, the plane began to oscillate violently. Any attempts by the autopilot to rectify matters only made things worse. The plane shook itself to pieces and broke itself up in midair before the crew could react.

'The plane shook itself to pieces and broke itself up in midair before the crew could react.'

8 September 1994

Flight: USAir Flight 427
Model: Boeing 737-3B7
Registration: N513AU
Fatalities: 132/132 (127 passengers and 5 crew)
Principle cause: control malfunction

The 737 was approaching the runway at Pittsburgh when it flew into the residual turbulence of the wake of a Delta Airlines 727, which had been in the same airspace 69 seconds before.

The 737 suddenly rolled 18 degrees left. When the autopilot reacted, it failed to level the plane out and the co-pilot initiated a compensatory right roll without disengaging the automatic system. The autopilot now tried to compensate for the co-pilot's input, and the plane rolled back violently to the left.

At 19:03:07 the plane was rolling left 70 degrees, and falling at a rate of 1100 metres per minute (3600 feet per minute). The plane stalled, then rolled upside down and nosedived. About a kilometre above the ground, it looked as if the plane was going to do a complete loop, but the nose dropped again and the plane went back into another dive. The 737 continued to roll, but the nose began to rise. At 2000 feet (610 metres) above the ground, the left roll hesitated briefly, then continued and the nose dropped again. The plane descended fast and hit the ground nose first at a speed of 483 kilometres per hour (300 miles per hour).

31 October 1994

Flight: American Eagle Flight 4184
Model: Avions de Transport Regional (ATR) 72-212
Registration: N401-AM
Fatalities: 68/68 (64 passengers and 4 crew)
Principle cause: ice

This was an accident notable not for the quantity of the fatalities, but for the horrific way in which the passengers died.

Flight 4184 was descending and approaching Chicago Airport. The plane's de-icing systems had already activated twice, when at 15:57:33 the ailerons suddenly deflected to a right wing-down position and the autopilot disconnected. The plane rolled left almost level, before suddenly rolling right again and continued to roll until the plane was upside down. Twelve seconds later the plane completed its roll.

The plane was stable and level again for three seconds, then at 15:57:48 the plane rolled left before going through another right roll. By now the nose was down 60 degrees, the airspeed was 485 kilometres per hour (300 miles per hour) and the occupants were experiencing more than twice the force of gravity. The crew still had some hope of regaining control but, by 15:57:55, the occupants were experiencing forces of over 3 G which soon increased to 3.6 G. In effect the passengers felt three times heavier than normal. Their own bodyweight was crushing them to death.

Data recording stopped when the plane was at 1682 feet (513 metres) with a vertical speed of 584 kilometres per hour (341 miles per hour). At 15:59 after 90 seconds of hell, the plane crashed nose-down, partially inverted into a soybean field near Roselawn, Indiana.

A TRAGIC MILESTONE

The UNITA accident remains the worse in Angola's history and the worst of all the hull losses of Lockheed L-188s.

18 December 1995

Flight: Trans Service Airlift Charter Flight
Model: Lockheed L-188C Electra
Registration: 9Q-CCR
Fatalities: 144/147 (137 out of the 139 passengers and 4 out of the 5 crew)
Principle cause: stupidity

Angola's UNITA movement had hired the Electra to fly some of its members into yet another war when, shortly after take-off, the plane crashed near Kahengula, Angola. In the depressing tradition of African aviation, the plane was carrying about 40 passengers beyond its capacity. It's amazing it managed to take off at all.

20 December 1995

Flight: American Airlines Flight 965
Model: Boeing 757-223
Registration: N651AA
Fatalities: 160/164 (152 of 156 passengers and 8 crew)
Principle cause: navigation failure and crew disorientation

The city of Cali lies high in the Andes, in a long valley between two rows of mountains 4300 metres (14 100 feet) high. The standard approach from the north, especially at night and in the dark, is with the aid of navigational waypoints (radar beacons that define the path to the landing strip or runway), specifically the Tulua Waypoint and the Rozo Waypoint.

Flight 495 was carrying mostly Colombians on their way to visit their families for the Christmas season. It had left Miami late at 18:35 and at 21:20 was 100 kilometres (60 miles) from Cali at a height of 36 000 feet (11 000 metres), approaching the Tulua Waypoint. Rebels opposed to the Colombian government had blown up the Cali radar, so there was no back-up to the onboard flight management system (FMS). ATC asked

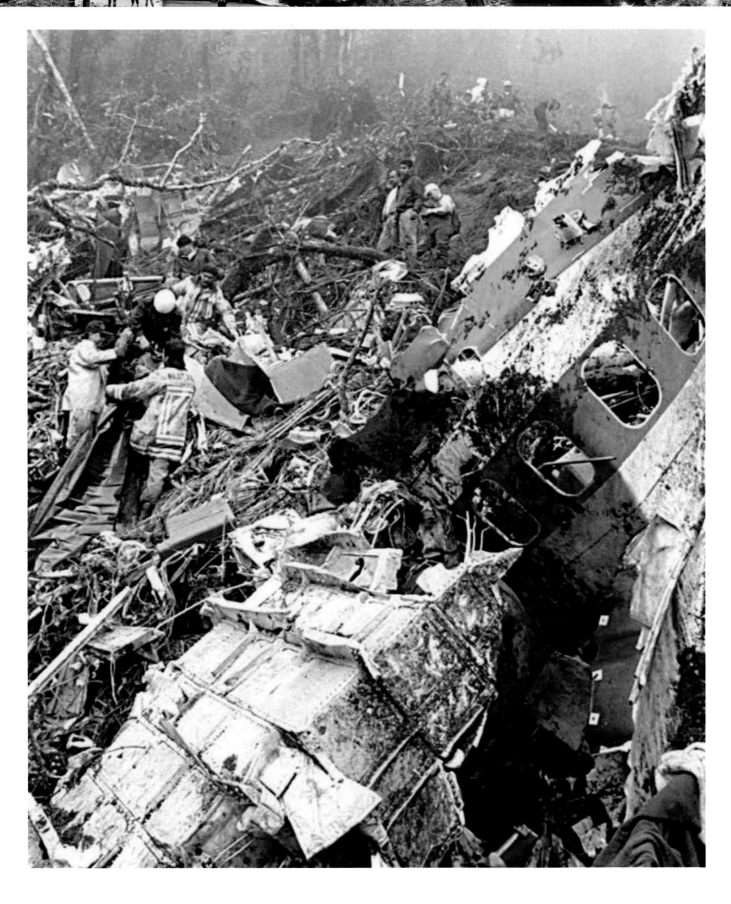

That investigators can glean so much information from such a mess seems rather miraculous to a layperson, and it's not easy even when you know how. The remains of American Airlines Boeing 757 which crashed into Mount El Deluvio, near Buga, Colombia, 20 December 1995.

Landings are statistically the most dangerous times during a flight. Some 45 per cent of all accidents occur on final approach and landing, far more than at any other time.

Flight 965 if they wanted to make a 'direct' approach to Cali from the north, rather than having to spend an extra twenty minutes looping around the airport to land from a southern approach. Flight 965 was already late and agreed.

This is where things went wrong. Captain Nicholas Tafuri misinterpreted the word 'direct' and cleared the waypoint data from the FMS assuming that he could program Cali directly. ATC then modified their instruction to 'report to Tulua'. Captain Tafuri and Co-Pilot Donny Williams now had to find Tulua manually. To speed up the plane's descent, they deployed the spoilers.

By the time they did find Tulua, they'd already passed it and had to program Rozo waypoint. To expedite matters Tafuri resorted to the FMS and punched in 'R' for Rozo. Usually the system automatically showed the closest 'R' at the top of the list. Unfortunately, since Colombia had two 'R' waypoints, the system defaulted to the largest city, Romeo, rather than the closest. When the captain pressed the 'execute' button, the plane's auto-guidance systems steered the plant to the east, towards Romeo.

The spoilers were still deployed and all the while the plane was travelling at 520 kilometres per hour (323 miles per hour) and descending at a rate of 1312 feet (400 metres) per minute.

By the time they realised their mistake it was night and pitch black. The crew reprogrammed the system to pinpoint on Cali's VHF radio frequency, but they were much lower down and in the next valley, so between them and the city was a wall of mountains that they couldn't see.

At 21:38 the GPWS activated. Immediately the pilots reacted to the 'Pull Up!' command, but the spoilers were still deployed. They were now fighting the machines that they had set and the plane couldn't gain enough altitude in time. Later research established that the plane would have made it had the spoilers been down.

The plane crashed into the top of El Deluvio Mountain and came to rest on the western side, 56 kilometres (35 miles) from Cali and far from any roads. It wasn't until 5 am the next morning, seven hours after the crash,

that there was enough light to send a rescue helicopter. Low cloud cover further hampered the search.

There were only four survivors. It was Mercedes Ramirez's 21st birthday and she lost both her parents in the crash. Nineteen-year-old Maurizio Reyez was able to rescue her. Gonzalo Dussan was able to save his daughter Michelle, but his son, Gonzalito died, trapped in a tree and suffering from massive internal injuries. All four had been sitting in seats next to the right wing.

American Airlines settled with the families of the victims then sued the designers of the navigation system. In June 2005 the Southern District of Florida Miami court found that American Airlines was 75 per cent responsible, while the navigation system designers were 25 per cent at fault.

6 February 1996

Flight: Alas Nacionales (ALW) Birginair Flight 301
Model: Boeing 757-225
Registration: TC-GEN
Fatalities: 189/189 (176 passengers and 13 crew)
Principle cause: faulty airspeed indicator and pilot error

Shortly after take-off from Puerto Plata in the Dominican Republic, the captain noticed that the airspeed indicator (ASI) wasn't working properly and reading an overspeed although the co-pilot's ASI seemed to be functioning. The autopilot reacted to the captain's readings, altered pitch and reduced power to lower the airspeed. Now both pilots were confused, since on the one hand the plane was reacting to overspeeding, while the co-pilot's ASI was reading lower.

Critically, the pilots both initially ignored the stick-shaker stall warning, which was in closest contact with reality. By the time they realised their error and turned off the autopilot, the plane had lost so much airspeed and altitude that it crashed into the ocean 26 kilometres (16 miles) north of Puero Plata, killing everyone onboard.

17 July 1996

Flight: Trans World Airlines (TWA) Flight 800
Model: Boeing 747-131
Registration: N93119
Fatalities: 230/230 (212 passengers and 18 crew)
Principle cause: centreline fuel explosion

FAMOUS CASUALTIES

Among those killed were French country and western guitarist Marcel Dadi, French artist Sylvain Delange and German–American photographer Rico Puhlmann.

At 20:31:12—the moment the CVR recording stopped on TWA Flight 800—the crew of an Eastwing Airlines Boeing 737 reported seeing an explosion. Six hundred and seventy eyewitnesses reported seeing the aeroplane breaking up and falling 4200 metres (13 800 feet) into the sea off east Moriches, New York.

Initially no one could rule out sabotage—the crash had happened only two days before the Atlanta Olympic Games—and some witnesses even reported seeing a missile bringing the plane down.

However, according to the official report, a short circuit outside of the centre wing fuel tank caused a spark in the fuel level indication system. This, in turn, ignited the centre-wing fuel tank.

The explosion was so violent that the front of the 747, including the cockpit and the entire first class section just fell off the plane. Without the weight of the front section, the plane climbed a further 3000 feet (900 metres) before diving into the ocean.

The Reverend Mychal Judge, Chaplain of the New York Fire Department, was on duty that day and for years after was a counsellor to the mourners. On 17 July 2001 he joined the relatives and friends for the fifth anniversary remembrance ceremony. 'Every year that you come here, you make this spot more blessed and more sacred,' Judge said. Six weeks later he would be dead, the first official victim of the 9/11 attack in New York.

The rumours remain even today. One witness reported seeing a reddish streak rising into the sky before reaching a white light moving across the sky, and that the explosion happened when the two met. A vocal group of conspiracy theorists, including some of the victims' family members, continue to contend that the American Government is covering up a 'friendly fire' scandal and that US Navy submarines conducting secret tests in the area at the time had made a fatal error.

The truth may never be known for certain. The official report cites the damage to the recovered plane fragments as proof that a missile did not hit the plane, but it nevertheless chooses its language to reflect its uncertainty of what did cause the disaster.

It seems a remarkably unconvincing explanation, considering the time and trouble involved. TWA 800 turned out to be the most intensive and the longest investigation in the NTSB's history.

The final report wasn't released until over four years after the accident. It was also the most expensive investigation of an aeroplane accident ever undertaken and is estimated to have cost around $120 million. There are still no industry-wide measures in place to ensure that the cause that officially brought down TWA 800 won't happen again.

The remains of the tail section of TWA Flight 800. The investigation of this particular crash has been the most expensive in history to date and perhaps the least convincing.

29 August 1996

Flight: Vnokovo Airlines Flight 2801
Model: Tupolev 154M
Registration: RA-85621
Fatalities: 141/141 (130 passengers and 11 crew)
Principle cause: translation problems

Although conversations between flight crew and ATCs are supposed to be conducted in English, fluency in that language varies greatly among pilots and control staff. On some occasions staff have to resort to native languages to make their points.

This seems to be the main problem with Flight 2801. The Russian plane was a charter flight, taking miners and their families from Moscow to mines near Svalbard on a mountainous island of Norway deep in the Arctic. Although the airport at Svalbard is only 27 metres above sea level it's in fjord country and surrounded by glaciers and mountains. As the plane was making its final approach, Svalbard ATC gave the Russian crew detailed instructions, but the Russians don't seem to have understood them. Throughout the final phase of the flight the plane remained consistently parallel, 3.7 kilometres (2.3 miles) to the right of the centreline that they should have been following.

The tail section of Korean Air Boeing 747 that crashed 06 August 1997 on the American occupied Pacific island of Guam. Twenty six of the 231 people aboard survived the tragedy.

As a result at 10:22 the plane's GPWS warning system sounded. Although the crew reacted by pulling up, the plane crashed nine seconds later into the top of Mount Operafjellet, at an altitude of 907 metres (2975 feet), still 3.7 kilometres to the right of the line that they should have been following. The accident remains Norway's worst ever air disaster.

11 November 1996

Flight: Aviation Development Corporation (ADC) Flight 086
Model: Boeing 727-231
Registration: 5NBBG
Fatalities: 143/143 (134 passengers and 9 crew)
Principle cause: air traffic control error

Flight 086 was on an approach to Lagos Airport, Nigeria. At the same time a Triax plane was departing Lagos on its way to Enugu, Nigeria. Although the ATC controller thought that he had cleared Flight 086 to descend to 10 000 feet (3000 metres), he hadn't and so cleared Flight 086 to descend a little late. This put it in the path of the of the Triax plane. The traffic collision avoidance system (TCAS) system sounded, but when the 727 pilots took evasive action they overcompensated, and the jet rolled too far. The pilots lost control and within sixteen seconds the plane was travelling upside down at 1200 kilometres per hour (760 miles per hour)—almost the speed of sound. The 727 disintegrated on impact at near Ejirin at 17:05.

6 August 1997

Flight: Korean Air Lines Flight 801
Model: Boeing 747-35B
Registration: HL7468
Fatalities: 228/253 (206 out of the 231 passengers and 22 out of the 23 crew)
Principle cause: controlled flight into terrain

The island of Guam, famous for its strategic importance and the battle that took place there during World War II, lies in the west Pacific and is the largest and southernmost island in the Mariana Archipelago. It's a popular holiday destination for Koreans and Japanese.

By the time that Korean Airlines 801 had travelled from Kimpo Airport in Seoul and was on its final approach, the plane had been in the air for about five hours and the landing at Agana, Guam, should have been routine.

Investigators later determined that a number of different factors contributed to what happened next. Under normal circumstances the plane would have used the VHS radio-based instrument landing system (ILS). However, on that particular day this highly dependable method was undergoing an upgrade, so the pilots had to use VHF omni-directional radio-range/distance measuring equipment (VOR/DME). This system is not quite as automatic as ILS. Crucially, the crew had to pay attention to their altitude.

The Agana Centre Radar Approach Control had a minimum safe altitude warning system (MSAW) installed in order to monitor the altitude of approaching aircraft and to alert all concerned if, for whatever reason, a plane was coming in too low. Unfortunately, the MSAW had recently had a software update that had caused so many false alarms that the system was useless. In the spirit of 'The Boy Who Cried Wolf' a programmer disabled the system. Even if it had been on, the system wouldn't have worked. In either case Agana Control failed to notice that Korean Airlines 801 was flying too low.

Nimitz Hill is in a direct line to the approach to the airport and Agana and reaches a height of 709 feet (216 metres). The 747 should have been flying at 14 500 feet (4420 metres). Instead the crew had failed to implement a proper approach and Korean Airlines 801 was flying at 650 feet (200 metres) when it hit Nimitz Hill at 01:42. As it crashed the plane hit an oil pipeline that feeds a power plant in Agana. Although the subsequent oil spillage failed to ignite, the fuel the plane was carrying was enough to turn the fuselage into a holocaust with wreckage strewn over the tropical jungle 5 kilometres (3 miles) away from the runway. The main body of the plane broke into four pieces, and the majority of the survivors came from the tail, which had remained intact.

Wreckage had trapped Seong Yeo Cho near the rear of the jet. She insisted that her eleven-year-old daughter, Rika Matsuda, sitting next to her, escape. Rika managed to crawl out of the plane before her mother died in the ensuing blaze and explosions. The Governor of Guam, Carl Gutierrez was involved hands-on in the rescue and eventually found Riki, who was lying in an exhausted sleep next to another survivor, flight attendant Yong Ho Lee.

Amazingly, the ATC transcript shows that it took the controllers at the airport around seven to eight minutes to realise what had happened. They didn't call emergency services until 25 minutes after the crash, assuming that the plane had aborted the initial landing and was making another approach.

A TRAGIC MILESTONE

Flight 152 remains Indonesia's worst ever plane disaster.

In the immediate aftermath over 500 rescuers worked at the crash site. The costs involving the US Navy alone came to almost $700 000. Rescuers took 32 survivors from the crash site, but two were dead on arrival—four others died over the next few weeks.

On 13 August 1997, a week after the crash and in the wake of delays in repatriating the bodies of the victims, 50 family members of the dead staged a much publicised sit-in protest at Agana's airport terminal.

Three weeks after the crash investigators had still not accounted for 23 bodies. They had, however, recovered 205 bodies, but had positively identified only 93. A black marble obelisk with the names of the victims inscribed now marks the place where over 200 people died.

In the wake of the debacle, the Federal Aviation Authority took a closer look at the minimum safe altitude warning system and found serious problems at over 60 other airports throughout the world.

26 September 1997

Flight: Garuda Flight 152
Model: Airbus A300B4-220
Registration: PK-GAI
Fatalities: 234/234 (222 passengers and 12 crew)
Principle cause: low visibility and miscommunication

Smog from recent forest fires enshrouded the region around Medan Airport, Indonesia, as Flight 152 was making its final approach. Visibility was down to 600 to 800 metres (2000 to 2700 feet). At 13:00, when the plane was at 3000 feet (900 metres) ATC ordered the plane to turn left and descend to 2000 feet (600 metres), but instead the crew confirmed a right turn. Sixteen seconds later the 727 crashed into a wooded area 32 kilometres (20 miles) south of Medan, near the village of Pancur Batu, between 900 and 1000 metres (3000 to 3300 feet) above sea level.

'In the spirit of "the boy who cried wolf"
      a programmer disabled the system. Even if it had been on,
                        the system wouldn't have worked.'

31 October 1999

Flight: EgyptAir Flight 990
Model: Boeing 767-366ER
Registration: SU-GAP
Fatalities: 217/217 (202 passengers and 15 crew)
Principle cause: suicide

At 01:19 Flight 990 left New York JFK bound for Cairo, Egypt. On board were two flight crews. At around

01:40 the relief first officer (RFO) suggested relieving the command first officer (CFO). The CFO agreed and left the flight deck. Eight minutes later the Command Captain (CC) left the flight deck to go to the toilet. Twenty-one seconds later the RFO, alone in the cockpit said quietly, 'I rely on God'—an unusual comment considering that there was nothing wrong at the time. What followed was even more unusual.

01:49:18 the RFO disconnected the autopilot.
01:49:48 the RFO repeated 'I rely on God'.

The pain of the loved ones isn't enough to stop the curiosity of the media or, by extension, our own.

The trauma of losing someone to a plane crash is something that can take years to recover from—if indeed you ever do.

01:49:54 the RFO initiated an abrupt nose-down elevator movement and pitches the plane into a rapid, nose down descent.

01:49:57 to 01:50:05 the RFO continued to quietly repeat, 'I rely on God', seven further times and the plane's descent caused the occupants of the plane to experience a noticeable weightlessness. The CC then re-entered the cockpit and asked loudly, 'What's happening? What's happening?' and the RFO repeated for the tenth time 'I rely on God.'

01:50:08 the 767 was now travelling faster than its maximum operating airspeed and a master warning alarm sounded as the RFO repeated, 'I rely on God'—to which all the CC could do was ask again, 'What's happening?'

01:50:24 the RFO had now placed the right and left elevators in opposite position, the engines to full throttles and the speed brake to full.

01:50:26 the CC, either still couldn't understand or believe it. He asked, 'Did you shut the engines?' and the RFO replied 'They're shut.'

01:50:31 to 01:50:37 the CC repeatedly ordered the RFO to 'Pull with me', but the controls remained unchanged.

01:50:38 the CVR and FDR stopped recording, but later analysis showed that the 767 climbed again to 25 000 feet (7620 metres) and changed heading from 80 degrees to 140 degrees. It then started a second descent and crashed into the ocean 100 kilometres (63 miles) south of Nantucket Island, Massachusetts.

30 January 2000

Flight: Kenya Airways Flight 431
Model: Airbus A310-340
Registration: 5Y-BEN, *Harambee Star*
Fatalities: 169/179 (159 of 169 passengers and 10 crew)
Principle cause: pilot error

The *Harambee Star* had just taken off Abidjan, Cote d'Ivoire, when, at 21:08:59 and at a height of 300 feet (90 metres), the stall warning sounded. In reaction,

the pilot pushed forwards on the control column and lowered the pitch, but failed to activate take-off/go around mode (TOGA). Without the extra thrust that TOGA would have provided, the pilot inadvertently put the plane into an accelerated descent. It took only 25 seconds for the plane to crash into the sea south of the airport. The remains of the Airbus sank in 40 metres (130 feet) of water.

19 April 2000

Flight: Air Philippines Flight 541
Model: Boeing 737-2H4
Registration: RP-C3010
Fatalities: 131/131 (124 passengers and 7 crew)
Principle cause: pilot error

Flight 541 had originated in Manila and was on the final approach to Davao on Samal Island in the Philippines. For reasons that remain unclear, the plane was flying at around 500 feet (150 metres) when it should have been flying at 1500 feet (460 metres). At 07:00 the 737 crash-landed into a coconut plantation, where it instantly caught fire and disintegrated. Although the weather was clear, the cloud ceiling over Davao was low and the pilot may have overcompensated.

23 August 2000

Flight: Gulf Air Flight 072
Model: Airbus A320-212
Registration: A4OEK
Fatalities: 143/143 (135 passengers and 8 crew)
Principle cause: pilot error

The official report cited principally that the pilots' failure to adhere to standard operational procedures led to this accident. The plane's flight data recorders weren't functioning properly, and this precluded finer analysis.

What is known is that Flight 072 from Cairo was making its final approach to Bahrain. When the plane was about 2 kilometres (1 mile) from the airport, the crew requested a left-hand orbit. This 360-degree loop was meant to slow the plane down and lower its altitude. When this didn't work well enough, the crew aborted the landing, but somehow, on the subsequent turn around and at a height of 1000 feet (300 metres) the Airbus started angling down. At 19:30 it pitched into the sea at a speed of 520 kilometres per hour (320 miles per hour) off Manama.

4 July 2001

Flight: Vladivostokavia Flight 352
Model: Tupolev 154
Registration: RA-85845
Fatalities: 145/145 (136 passengers and 9 crew)
Principle cause: pilot error

The Tupolev had originated in Yekaterinburg and was on its final approach to Irkutsk, Russia. The co-pilot was flying the plane at 2640 feet (805 metres) when the plane entered a flat spin, which continued for 22 seconds as the plane descended. Although the pilot applied full throttle to escape the spin, he failed to gain control and the plane crashed, belly down at 02:10 near Burdakovka.

12 November 2001

Flight: American Airlines Flight 587
Model: Airbus A300B4-605R
Registration: N14053
Fatalities: 260/260 (251 passengers and 9 crew) and 5 on the ground, total 265
Principle cause: pilot error

Barely two months after 9/11, the citizens of the New York City borough of Queens must have thought that the nightmare was starting all over again.

Flight 587 took off from New York JFK at 09:14 bound for Santo Domingo and following in the wake of a JAL 747 that had preceded it about one minute and 45 seconds earlier. One minute and fifteen seconds into the flight, the plane encountered some wave turbulence and it began to yaw right. The co-pilot who was flying the plane then applied full right and left rudder at maximum power—a move that the accident inquiry would later regard as 'unnecessary and excessive'. The manoeuvre stressed the tail assembly beyond its manufacturing tolerances, and at 09:16, when the plane was executing a left turn, the entire vertical tail fin snapped off.

Witnesses reported seeing the plane's number one engine catch fire, then falling debris from the left wing. The number one engine then separated from the wing. Shortly after the right number two engine fell off. Both engines landed within 30 metres of each other near the corner of Newport Avenue and Beach 129th Street. Engineless and rudderless, the doomed Airbus dived and crashed into Beach 131st Street near Rockaway Beach, a residential district of Queens. The crash destroyed four houses, severely damaged eight others and killed five people on the ground.

Investigators found the tail fin over 3 kilometres (2 miles) away in Jamaica Bay.

15 April 2002

Flight: Air China Flight 129
Model: Boeing 767-2J6ER
Registration: B-2552
Fatalities: 129/166 (121 out of 155 passengers and 8 out of 11 crew)
Principle cause: pilot error

Flight 129 approached Pusan Airport in South Korea at the end of its journey from Beijing, China. Although the 767 was originally meant to land on runway 36L, ATC changed the plan to runway 18, requiring a new approach, which none of the pilots on Flight 129 had flown before.

Crashes in highly populated areas are comparatively rare, but when they do happen the results make even bigger headlines. When an American Airlines Airbus A300 crashed into the suburb of Belle Harbor in Queens, New York, barely two months after the World Trade Centre attack, Americans thought the nightmare was starting all over again.

A Philippine Coast Guard diver retrieves a body from a Fokker 27 plane that crashed into Manila Bay shortly after takeoff 11 November 2002.

As a result of poor visibility and the unfamiliarity of the approach, the crew lost their bearings and, when they lost sight of the runway, they did not immediately execute a standard 'missed approach' manoeuvre. In the confusion the 767 hit a mountain 5 kilometres (3 miles) north of the airport.

25 May 2002

Flight: China Airlines Flight 611
Model: Boeing 747-209B
Registration: B-18255
Fatalities: 225/225 (206 passengers and 19 crew)
Principle cause: structural collapse

On 7 February 1980 the plane's tail had hit the runway in Hong Kong. Subsequent investigation revealed that maintenance teams had not repaired the plane in accordance with Boeing guidelines.

Over twenty years later the plane had made over 20 000 trips, which had gradually fatigued the compromised hull. Flight 611 took off from Taipei Airport at 15:07, bound for Hong Kong. At 15:30 the plane was at 35 000 feet (10 670 metres), when it broke up in midair and disappeared from radar screens. The main wreckage fell into the Taiwan Strait 45 kilometres (28 miles) north-east of the Taiwanese Islands of Penghu. Some pieces of the plane ended up as far as Changhua, 45 kilometres (28 miles) from the main crash site.

The 23-year-old plane had been on its last flight before China Airlines was about to sell it to Orient Thai Airlines. As it was, Orient Thai were lucky, which is more than could be said for the 225 who died on Flight 611. It's a miracle that the plane flew as long as it did.

Taiwan immediately grounded China Air's four other 747s before eventually clearing them.

19 February 2003

Flight: Islamic Revolution's Guards Corps
Model: Ilyushin 76MD
Registration: 15-22
Fatalities: 275/275 (257 passengers and 18 crew)
Principle cause: poor weather

Although a military accident, the loss of life was so great that at the time of writing this is the ninth worst air disaster in history. At 18:24, in poor, winter weather and 35 kilometres south-east of its destination of Kerman, Iran, the Ilyushin crashed 90 metres (300 feet) below the summit of a 3500 metre (11 500 foot) high peak in the Sirach Mountains.

25 December 2003

Flight: Union de Transports Aériens de Guinee Flight 141
Model: Boeing 727-223
Registration: 3X-GDO
Fatalities: 141/163 (136 of 153 passengers and 5 of 10 crew)
Principle cause: criminal negligence

It's not just how much a plane weighs that's important. It's also how that weight is distributed in the cargo hold. Weight distribution is a specialist skill that competent carriers take very seriously.

Investigators into this accident later concluded that:

'The operator's serious lack of competence, organisation and regulatory documentation, which made it impossible for it both to organise the operation of the route correctly and to check the loading of the aeroplane ...' and 'The inadequacy of the supervision exercised by the Guinean civil aviation authorities and, previously, by the authorities in Swaziland, in the context of safety oversight ... all contributed to this disaster.'

'The 23-year-old plane had been on its last flight before China Airlines was about to sell it to Orient Thai Airlines. As it was, Orient Thai were lucky, which is more than could be said for the 225 who died on Flight 611.'

On Christmas Day Flight 141 was overloaded and front heavy when it started its take-off from the runway at Cotonou Airport in Benin. By the time it took off, the plane barely climbed and its unretracted landing gear hit some of the airport's localiser antennas. The plane then struck the roof of a radio building, injuring the occupant, before smashing through the concrete airport boundary fence and breaking up on the beach next to the airport.

3 January 2004

Flight: Flash Airlines Flight 604
Model: Boeing 737-3QB
Registration: SU-ZCF
Fatalities: 148/148 (135 passengers and 13 crew)
Principle cause: undetermined crash out of control

Flight 604 was departing Sharm el Sheikh-Ophira and on its way to Cairo when, at an altitude of 5460 feet (1665 metres), and for reasons that remain unclear, the plane began to bank to the right. Although the crew managed a slight recovery, the plane continued to gain speed and lose height. At 04:45 the plane struck water at a speed of 770 kilometres per hour (479 miles per hour). The g forces in the plane were 3.9, so the occupants subjectively felt almost four times their normal weight. What was left of the plane and the people on it sank in water 900 metres (3000 feet) deep.

16 August 2005

Flight: West Caribbean Airlines Flight 708
Model: McDonnell Douglas MD-82
Registration: HK-4374X
Fatalities: 160/160 (152 passengers and 8 crew)
Principle cause: mechanical failure/pilot error

Flight 7080 was en route from Panama City to Fort de France, Martinique, when at 01:49, nearly 50 minutes into its journey and at a height of 33 000 feet (10 000 metres) the plane experienced engine trouble. The crew diverted to Caracas.

At 02:57 the plane continued to lose power, both engines flamed out, airspeed slowed and the plane began to fall. At 25 000 feet (7620 metres) airspeed dropped to 278 kilometres per hour (173 miles per hour) and the plane stalled.

At this point the nose of the plane pitched up, and instead of pushing forwards to lower the nose, increase airspeed and escape the stall, it seems that the pilots may have panicked and they continued to hold their control yokes to their chests.

The plane continued to descend without power at a rate of 7000 feet (2133 metres) per minute and eventually crashed into a swamp in the Sierra de Periha Mountains on the border of Colombia and Venezuela.

The data recorder noted that the pilots held their yokes to their chests all the way to the ground.

The wreckage of a Boeing 727—full of Lebanese expatriates heading home for Christmas—that crashed on take-off 25 December in Benin. Twenty two of the 163 people aboard survived the crash.

'The data recorder noted
that the pilots held their yokes
to their chests all the way to the ground.'

9 July 2006

Flight: Sibir (S7) Airlines Flight 778
Model: Airbus A310
Registration: 324
Fatalities: 125/203 (120 out of 195 passengers and
5 out of 8 crew)
Principle cause: pilot error

At 07:50 Flight 778 had come from Moscow and just touched down on the runway at Irkutsk Airport, Russia. While it was slowing down the pilot accidentally activated the number one engine power lever, and at a critical point there was forwards thrust on the left with reverse thrust on the right. The result was that the plane ran off the runway, and, even though it was only travelling at a comparatively modest 80 kilometres per hour (50 miles per hour), the plane crumpled and burst into flames when it hit a concrete barrier.

MIDAIR COLLISIONS SEEM THE most unlikely of events. You'd think that there would be a lot of room for planes to manoeuvre, but they usually fly along air corridors. Like roads or highways in the sky, these well-defined flight paths ensure the planes stay on course, are within range of navigational aids and restrict the environmental impact on areas over which they fly.

Mistakes do happen, however. Planes may stray from their designated flight paths, there may be communication problems between air traffic controllers and sometimes

PART FOUR:
Bizarre and unusual accidents of the Jet Age

04

two planes just happen to be in the wrong place at the wrong time. Because midair collisions involve more than one plane and occasionally occur over populated areas, the total death toll is often greater that those accidents involving only single aircraft.

While midair collisions are unusual, they are by no means the freakiest of accidents. Sometimes the chain of events that brings a plane down is, frankly, so weird that you begin to feel the guiding hand of fate. In some cases 'accident' seems too inadequate a term.

18 May 1935

Flights: Gor'ky Eskadril'ya Demonstration Flight
and Russian Airforce I-5
Models: Tupolev ANT-20
Registrations: CCCP120, *Maxim Gorky*
Fatalities: 50/50 (37 passengers and 13 crew)
and 1/1 (pilot), total 51
Principle cause: midair collision

The *Maxim Gorky* was on its maiden flight, and the I-5
was performing acrobatic loops around the Tupolev at
an altitude of 2300 feet (700 metres) when it collided
with the passenger plane at 12:45. The planes crashed
into a residential neighbourhood near Moscow, but there
were no recorded ground casualties. It was the worst air
accident of the year.

1 November 1949

Flights: Eastern Air Lines Flight 537 and Bolivian
Air Force
Models: Douglas C-54B-10-DO and P-38
Registrations: N88727
Fatalities: 55/55 (51 passengers and 4 crew)
and none
Principle cause: midair collision

Flight 537 was cleared to land at Washington National
Airport. ATC instructed the P-38 behind it to land
after it. The P-38 failed to confirm or comply with the
instruction and continued its descent. ATC instructed
Flight 537 to turn left, and, at 11:46, about a kilometre
(0.5 miles) short of the runway both aircraft collided
at an altitude of 300 feet (90 metres), killing everyone
onboard the passenger plane. The P-38 pilot survived
with serious injuries.

25 April 1951

Flights: Cubana de Aviacion Flight 493 and US Navy
Models: Douglas DC-4 and Beech SNB-1 Kansan
Registrations: CU-T188 and 39939
Fatalities: 39/39 (34 passengers and 5 crew)
and 4/4 (crew), total 43
Principle cause: midair collision

Both planes were operating under visual flight rules
and the subsequent investigation determined that
both crews failed to 'maintain sufficient vigilance'.
The commercial plane was flying from Miami to
Havana, Cuba, when at 11:49 it crashed into a military
plane on a training flight. The SNB immediately fell 4000
feet (1200 metres) and crashed into the waters off Key
West, Florida, just west of its base. The DC-4 banked
left and lost altitude, until it plummeted into the ocean
about 3 kilometres (2 miles) from where the collision
had taken place.

30 June 1956

Flights: Trans World Airlines Flight 2 and United
Airlines Flight 718
Models: Lockheed L-1049 Super Constellation
and Douglas DC-7
Registrations: N6902C, *The Star of the Seine*,
and N6324C
Fatalities: 70/70 (64 passengers and 6 crew)
and 4/4, total 74
Principle cause: midair collision

In accordance with their flight plans these planes should
have had 300 metres (1000 feet) of air separating them.
The official investigation could not determine why the
crews of both planes failed to see each other, but it cited
the following possibilities:

When planes collide it is often surprising what remains relatively intact. Engines are usually the toughest parts, unless they explode.

The Lockheed Super Constellation and its sister model the Stratoliner, were the ultimate planes before the jet age. Propeller-driven passenger planes would never get any better. Even they were not immune to midair collisions though.

1) Intervening clouds reducing time for visual separation
2) Visual limitations due to cockpit visibility
3) Preoccupation with normal cockpit duties
4) Preoccupation with providing passengers with a more scenic view of the Grand Canyon area
5) Physiological limits to human vision reducing the time opportunity to see and avoid the other aircraft
6) Insufficiency of en-route air traffic advisory information due to inadequacy of facilities and lack of personnel in air traffic control.

In short, the report concisely expresses nearly every potential cause for a midair collision. Whatever the true cause, we do know for certain that at 10:32 and at an altitude of about 20 000 feet (6000 metres), the centre fin leading edge of *The Star of the Seine* struck the left aileron tip of the DC-7. Then the lower surface of the left wing of the DC-7 crashed into the upper aft section of the *Star*'s fuselage. The left wing of the DC-7 broke off as it cut off the lower plane's tail. Both planes were mortally crippled and their crash into the Grand Canyon below was inevitable.

16 December 1960

Flights: United Airlines Flight 826 and Trans World Airlines Flight 266
Models: McDonnell Douglas DC-8-11 and Lockheed L-1049S Super Constellation
Registrations: N8013U and N6907C
Fatalities: 84/84 (77 passengers and 7 crew), 44/44 (39 passengers and 5 crew) and 5 on the ground, total 133
Principle cause: midair collision

TWA Flight 266 left Port Columbus Airport, Ohio, at 09:00 bound for New York-LaGuardia Airport. United Air Lines Flight 826 left Chicago O'Hare Airport at 09:11 bound for New York-Idlewild (later JFK) Airport.

At 10:30, as both planes were reaching their respective destinations, New York ATC cleared United 826 to descend from 14 000 feet to 5000 feet (4300 to 1550 metres), maintain the lower level in a holding pattern and to await final clearance for landing from the approach controller.

However, the crew of United 826 didn't know that they were already past the area where they were supposed to hold and approaching the Miller Army Air Station on Staten Island; they even confirmed to New York ATC that they were on their way to LaGuardia. Meanwhile TWA 266 was descending to 5000 feet and heading for Miller Station too.

At 10:33:14, LaGuardia Approach Control issued clearance to TWA 266 to continue their descent to 1500 feet (1600 metres) but advised them that there 'appears to be jet traffic off your right now 3 o'clock at one mile, north-east-bound.'

With both planes travelling at hundreds of kilometres per hour, there was barely enough time to react. Within seconds, at an altitude of 5200 feet, UA 826's number four engine sliced open the top of TWA 266, causing that plane to break up into three sections before crashing into Miller Army Air Station. UA 826 stayed in the air just long enough to travel another 13.5 kilometres (8.5 miles) in an attempt to make an emergency landing, before it crashed into a residential area of Brooklyn, killing everyone onboard as well as five other people on the ground.

12 December 1961

Flight: British European Airways Scheduled Flight
Model: de Havilland DH-106 Comet 4B
Registration: G-ARJM
Fatalities: 27/34 (20 of 27 passengers and all 7 crew)
Principle cause: instrument malfunction

This Comet was one of the more recent models, a 4B, and structurally superior to earlier comet models that had failed so spectacularly on account of metal fatigue some

The Comet—the plane that ushered in the new era of commercial passenger travel, but also became its first casualty.

twenty years before. Only a few seconds after take-off, the Comet assumed a very steep climbing angle. Witnesses reported a pitch as high as 45 or 50 degrees as well as wing drop and engine noise variations. The plane stalled at an altitude of just 450 feet (140 metres) with the left wing down. It then levelled off and simply dropped to the ground around 1600 metres (1 mile) south west of the control tower at Ankara-Esenboga Airport in Turkey.

Investigators determined that one of the three screws holding the face of the captain's horizon indicator had worked loose and jammed the indicator pointer. The pointer therefore gave an incorrect indication of pitch. One could argue that the captain gave greater weight to what his instruments were telling him than what his body was telling him. One assumes that in spite of G-forces and acceleration he would have been able to feel the real angle of the plane, but when the jet was taking off, there may not have been enough time to make a judgement call when the crew had to balance so many other factors.

23 November 1962

Flight: United Airlines Flight 297
Model: Vickers Viscount 745D
Registration: N7430
Fatalities: 18/18 (14 passengers and 4 crew)
Principle cause: striking a swan

At 12:24 while cruising at an altitude of 6000 feet (1800 metres) over Ellicott, Maryland, the plane flew into a flock of Whistling Swans. One weighing an estimated 6 kilograms (13 pounds) struck the leading edge of the left horizontal stabiliser and caused it to break off. The crew lost control and the plane broke up in midair.

'...the plane flew into a flock of Whistling Swans ...'

LIGHTNING STRIKES

There were cases of lightning or electrical discharges bringing airships down but, for reasons that remain obscure, the air industry thought that airliners, especially jet aircraft, were somehow immune to being struck by lightning. It took two separate accidents, four years apart, before authorities realised their mistake and acted.

26 June 1959

TWA Flight 891, a Lockheed L-1649A, was flying between Milan, Italy, and Paris, France. Just 32 kilometres north-west of its departure point and 15 minutes into its flight, the plane was climbing at an altitude of 10 000 feet (3050 metres) through stormy weather when a combination of static electrical discharges and lightning ignited fuel vapour in tanks six and seven. The subsequent explosions tore the wings off and 59 passengers and nine crew members died in the accident.

8 December 1963

Pan Am 214, a Boeing 707-121, the *Clipper Tradewind*, had left Friendship International Airport (now Baltimore-Washington International Airport) at 20:24. Unfortunately, extreme winds at its destination airport of Philadelphia forced the *Clipper* and five other aircraft into a holding pattern.

Later investigations concluded that, at 20:58, positive lightning struck the left wing. Although positive lightning comprises only 5 per cent of all lightning strikes, it is immensely powerful and can carry ten times as much current as normal negative lightning. The discharge ignited the left fuel tanks first, then almost instantly ignited the central and right fuel tanks. In the subsequent explosions, the left wingtip blew off completely. In about a minute the plane fell out of the sky.

The crew managed a 'Mayday' and a final message of 'Clipper out of control', before the plane crashed into the ground near Elkton, Maryland, killing all 81 people onboard (73 passengers and 8 crew).

In the aftermath the Federal Aviation Authority ordered that operators install lightning discharge wicks on all aircraft operating in American airspace. It retrospect it seems as if even Mother Nature had it in for Pan American Airlines.

5 March 1966

Flight: BOAC Flight 911
Model: Boeing 707-436
Registration: G-APFE
Fatalities: 124/124 (113 passengers and 11 crew)
Principle cause: turbulence

BOAC Flight 911 from Honolulu was supposed to have arrived at Tokyo on 4 March but had needed to stay overnight at Itazuke Airbase in nearby Fukuoka. After finally making it to Tokyo the next day, it took off again at 13:58. Shortly after take-off the captain decided to give his passengers a treat and fly near Mount Fuji. In unusually clear skies the view was spectacular, but while flying over Gotemba city at a height of (15 800 feet) (4900 metres) and with an airspeed between 600 and 685 kilometres per hour (370 to 425 miles per hour), the plane started trailing white vapour. Over the next few seconds the plane found itself in the midst of severe clear air turbulence. The abnormally high gust load was beyond the design limits of the 707 and it broke apart in midair. The forward fuselage broke off over Taroba and the rest of the plane crashed in a forest 300 metres west, where it caught fire. This case continues to claim the highest ever mortality in an accident in which experts attributed turbulence as the sole cause.

22 April 1966

Flight: American Flyers Airline Chartered Flight
Model: Lockheed L-188C Electra
Registration: N183H
Fatalities: 89/103 (83 out of the 98 passengers and all of the 5 crew)
Principle cause: heart failure

Reed Pigman was a dedicated pilot. So dedicated in fact that he was willing to falsify his medical records so that they made no mention of his heart condition

or his diabetes. When the plane he was flying missed its instrument landing he circled the plane to make a visual approach, but he had a heart attack and crashed the Lockheed into a hill near Ardmore, Oklahoma, at 20:30 and killed himself, his crew and most of the army personnel that had trusted him with their lives.

The official report referred to the cause as a 'coronary insufficiency'.

23 June 1967

Flight: Mohawk Airlines Flight 40
Model: BAC 111-204AF
Registration: N1116J
Fatalities: 34/34 (30 passengers and 4 crew)
Principle cause: incorrect installation

Engine bleed air (the air coming out of an engine) is very hot and supposed to flow in one direction only, so planes are fitted with a valve to ensure this, but what happens when you install the valve backwards? Unfortunately, this is what happened to this particular plane. The hot engine air blew into an air intake valve and through an auxiliary power unit, which heated the air up even more, so much so that sound-dampening materials caught alight. This fire then ignited hydraulic fluid in the plane's pitch controls. At 14:47, shortly after the BAC took off from Elmira Regional Airport in New York, observers near Blossberg noticed that large parts of the tail separated from the plane. Seconds later Flight 40 crashed to the ground.

5 July 1970

Flight: Air Canada Flight 621
Model: McDonnell Douglas DC-8-63
Registration: CF-TIW
Fatalities: 109/109 (100 passengers and 9 crew)
Principle cause: pilot error

On 1 December 1984 NASA deliberately crash landed a remote-controlled Boeing 720 in order to test a new fire-retardant fuel. As you can see, the experiment was somewhat less than an outstanding success, but it did allow observers to examine—in fine detail—the mechanics and utter devastation of a plane crash as it was actually happening. The 720 was filled with crash test dummies, none of which survived the crash. (See The Fuel Question on page 283.)

Two F-86 Sabre jet fighters take off. Not the sort of thing you want to meet in midair.

Anyone who has ever flown close to the wing of a commercial aircraft and who has been in an observant mood would have noticed that when the plane is landing two obvious things happen. Firstly the rear flaps on the wing angle sharply down. This is one of several methods aircrew use to slow the plane. The other thing is that thin flaps on the top of the wing angle up. These are the spoilers. They 'spoil' the airflow over the wing and disrupt the vacuum that keeps the plane up, so that the plane drops in a controlled stall—useful when you want to land. As a safety precaution crews have to arm the spoilers to prevent their accidental deployment while in the air. Spoiler deployment at the wrong time is not a good thing.

The crew of Flight 621 had agreed that when they were flying together they wouldn't arm spoilers until close to touch down. Although this was contrary to Air Canada policy, it made this particular pilot and first officer feel better.

Ironically, on the afternoon in question the first officer of Flight 621 deployed the spoilers too early, when the plane was still about 60 feet (20 metres) in the air. The plane dropped down too hard, too soon and lost its number four engine and some of the right wing lower plating. The exposed fuel feed to the missing engine was leaking fuel, which caught fire as the captain aborted the landing and climbed again. The first officer apologised to the pilot several times, but the plane was doomed. The pilot wanted to turn around and land directly on the same runway, but the fallen debris made that impossible.

ATC suggested another runway and the plane was in the air for another 150 seconds trying to reach the new runway, when there was an explosion. Parts of the right outer wing then broke off. Six seconds later another explosion ripped the number three engine from its mounting and a large section of the right wing along with it. The plane was now uncontrollable. More pieces of the right wing continued to break off before the plane crashed into the ground nose down, travelling at 410 kilometres per hour (255 miles per hour). The

subsequent explosion was so great that what remained of the aircraft and its occupants were fragments—none were larger that a few metres long.

30 July 1971

Flights: All Nippon Airways Flight 58 and Japan Air Self Defence Force
Models: Boeing 727-281 and F-86 F Fighter
Registrations: JA8329 and 92-7932
Fatalities: 162/162 (155 passengers and 7 crew)
Principle cause: midair collision/involuntary homicide

ANA 58 was en route from Sapporo to Tokyo at an altitude of 28 000 feet (8500 metres) when it entered airspace in which a student fighter pilot and his instructor were practising manoeuvres. At around 14:00 the instructor failed to warn the student of the approaching 727 in time. Although the pilot took evasive action and banked to the left, the right wing of fighter plane clipped the left horizontal stabiliser of the passenger jet. The 727 lost control and crashed near Shizukuishi in Iwate, Northern Honshu. The fighter plane's right wing broke off in the collision and crashed, but the pilot and his instructor ejected in time. Both were indicted for involuntary homicide. The instructor was convicted and sentenced to three years in jail for inadequate supervision of his student. The trainee pilot was acquitted.

DANGEROUS DOORS

Although it is not immediately apparent, doors—especially cargo doors—are one of the most vulnerable parts of an aircraft. Of particular concern is that doors have to withstand the stresses of internal pressure. Door-locking mechanisms can be complicated and door-locking procedures are not always straightforward. When doors fail explosive decompression follows—with devastating results. Doors have been central to several major and historically important air accidents and have revealed shortcomings in design, safety protocols and—most damningly—the power and willingness of authorities, manufacturers and operators to put profits and politics ahead of peoples' lives.

12 June 1972

American Airlines Flight 96

In 1967 aerospace manufacturer McDonnell bought Douglas Aircraft. Although Douglas had produced some important aeroplanes, experts thought it was now about a year away from bankruptcy. Initially the merger was a happy one and the new company expected the first product of its merger, the DC-10, to be even better.

The DC-10 was to be a wide-bodied carrier, which would compete with Boeing's 747, the Airbus A300 and the Lockheed L-1011 TriStar. However, at some point the new company seemed to falter and the founding companies that individually had had reputations for quality and safety soon lost a lot of goodwill.

As early as 1969, when McDonnell Douglas was designing the DC-10, engineers foresaw that there could be disaster if a particular sequence of events were to take place. Jet planes fly so high that they need pressurisation. Doors that only open inwards are good because internal cabin air pressure helps to keep the door sealed. However, you can fit more cargo in a hold if the door opens outwards, so designers have to make a judgement call. McDonnell Douglas opted for outward-opening doors. The designers of DC-10 doors, General Dynamics Convair Division did their best.

The original design of the DC-10 required that the baggage handlers do three things to shut the heavy cargo doors: pull down the top-hinging door; swing down a lever on the outside of the door; and press and hold a button to activate the closing mechanism hydraulics.

However, American Airlines—a major client of McDonnell Douglas—wanted the manufacturer to make the design lighter, so they complied and insisted that Convair use an electric motor for the closing mechanism instead. Convair didn't think this was safe and sent McDonnell Douglas a report warning of its vulnerability.

In May 1970, during a cabin pressure test, a deliberately badly closed door blew out. This depressurised the test plane's cargo hold. With the passenger cabin above now at a considerably higher pressure than the cargo hold, the floor of the cabin gave way.

McDonnell Douglas blamed Convair for a weak cabin floor and pressured the company into a quick fix. They fitted a vent flap onto the outer hull, which was meant to only close properly if the door closed properly. If the panel didn't close, the cabin would be impossible to pressurise, warning the flight crew that something was wrong. However, the designers didn't count on human psychology. If baggage handlers were having trouble closing the door, some would be inclined to force the issue. If the door was forced closed, the vent could still be shut, making a mockery of the whole failsafe. No one noticed this design flaw until it was too late.

Bryce McCormick was a 28-year veteran pilot with American Airlines who felt a tremendous responsibility for the people whose lives depended on his professional competence. When the Douglas DC-10 was first introduced into service in early 1972 he examined the

'Fortunately, Bryce invested the best few hours of his life working out how to solve the problem.'

new plane from top to tail. He noticed that although there were three independent cables and hydraulic lines controlling the tail section, the lines were all next to each other. If a problem disrupted one line, it was likely to disrupt the other two as well and there was no manual back-up if the hydraulics failed.

Fortunately, Bryce invested the best few hours of his life solving the problem. In the event of a hydraulic systems failure he found a way to land the DC-10 using only the plane's throttles—steering by powering one engine up, while powering another down—and by doing a similar thing with the tail-mounted engine to control the pitch of the plane, which let him climb or descend.

On the morning of AA Flight 96 and only two months after Bryce McCormick's self-training, the nightmare that Convair had predicted actually happened. Detroit baggage handler William Eggert had not closed the DC-10's cargo door properly—handlers had long been complaining that the DC-10 cargo door was difficult to

shut—and an electronic alarm had failed to notify the flight crew that anything was wrong.

As the plane took off and climbed, the relative air pressure between the interior of the DC-10 and the outside became greater and greater. Eventually the door pins began to shear. At an altitude of 12 000 feet (3600 metres) over Windsor, Ontario, the cargo door blew out.

The occupants heard a bang at the rear of the DC-10. In the cockpit there was a sudden depressurisation, which blew dust and rivets into the pilots' faces and knocked Bryce's headset off his head. The control columns became unusable and the plane went into a right-hand turn that was about to turn into a spin dive.

At the rear of the plane the cocktail bar collapsed through the floor and jammed the control cables leading to the tail. Bryce quickly forced the engines to full throttle, nipping the spin dive in the bud. He powered down the tail engine and, with incredible grace under pressure, he got what control he could of the plane.

The unique design of the DC-10 allows the pilot to control the aircraft—even after a major hydraulic failure—by using the throttles alone.

French officials survey the corpses of Turkish passengers who would not have died if aviation authorities had done their jobs properly.

Flight attendant Bea Copeland had almost fallen through the hole in the floor and with equal cool announced over the PA that they'd all be put on a new plane at Detroit on account of their 'mechanical problem'.

Bryce made it back to Detroit with the gentlest of throttle manoeuvres. As the plane started to land it was coming down at 300 kilometres per hour (186 miles per hour). It was way too fast, but he had no choice. With no hydraulic controls over the ailerons, the only way that he could keep the nose from dipping was to keep the speed up.

The DC touched down, started skidding and was about to crash into the airport fire station when Co-Pilot Paige Whitney came into his own. He cut back on the right thrust reverser and added power to the left in order to compensate for the plane's skewing to the right.

The plane ultimately came to a stop at the end of the runway, half on concrete, half on grass. Not one person died. The flight and landing of the crippled plane had been one of the most skilful examples of commercial piloting ever.

Bryce was to add his voice to the choruses demanding that they 'fix the damned door.'

DC-10 fixed the doors—sort of. They installed reinforcements on the doors of almost all DC-10s.

Crucially, they left one out. In 1974, in another bureaucratic blunder, somebody at McDonnell Douglas incorrectly approved a quality control report indicating that the company had fitted a Turkish DC-10 with an improved mechanism. This failure was to lead to what would become the worst civil aviation disaster ever up to that point.

3 March 1974

Flight: Turk Hava Yollari (THY) Flight 981

Model: McDonnell Douglas DC-10-10

Registration: TC-JAV

Fatalities: 346/346 (334 passengers and 12 crew)

Principle cause: faulty door

The THY flight had arrived from Istanbul with 117 passengers. At Paris, 217 others joined the flight. Many of them had originally wanted to fly British Airways, but a strike among ground staff at Heathrow Airport had forced them to change.

The DC-10 left Charles de Gaulle Airport and was travelling across France. When the plane reached 12 000 feet (3600 metres)—exactly the same altitude as AA 96 almost two years before—the cargo door blew out. The floor collapsed and six passengers were blown out the hole in the explosive decompression that followed.

The Turkish pilots were not prepared as Bryce McCormick had been. The plane that they were flying was heavy—fully loaded—and operating close to its limits. In an almost identical way to AA 96, the pilots lost hydraulic control of the systems. They struggled for 77 seconds while the plane went into a dive.

At 12:42, just 8 kilometres (5 miles) short of the runway, the DC-10 crashed into the forest of Ermonville. Everyone onboard died—more than twice as many as the previous record of 171 fatalities of a Royal Jordanian Airlines flight on 22 January 1973 at Kano, Nigeria.

The plane disintegrated into so many pieces that even today you can still find small, scattered remnants of that crash.

A joint team comprising the French Secretariat General Civil Aviation (SGAC) and the United States National Transportation Safety Board (NTSB) gradually unearthed a scandal.

Ever since the Ontario Crash, the NTSB—an independent organisation, like the Australian Air Transport Safety Bureau—had pushed the Federal

WORDS OF A WITNESS

In his book Darkest Hours, author Jay Robert Nash quotes from a French farmer who witnessed the disaster:

I didn't know it was a dead body ... I thought it was something else falling from the explosion ... It was a woman, although I could not be sure it was a woman because she was completely smashed, completely broken ... her head was here, her brains were here ... Then I ran to another one, and I saw the other woman ... one breast had been torn off. She was dead, too, and completely broken.

Aviation Authority in the America for a compulsory change to the DC-10 doors. For all its investigative power, the NTSB cannot legally force anyone to do anything; it's up to the FAA to issue directives to the air industry. In this case, the FAA didn't do enough.

In retrospect there seemed to be some collusion between the FAA and the plane manufacturer. McDonnell Douglas didn't want there to be any implication that the planes were not airworthy, and the modifications were expensive. The FAA seemed to be complacently expecting that whatever MD was doing was adequate. Furthermore, McDonnell Douglas expected the US government to fund any further research into better doors.

This attitude proved to be disastrous. With public revelation of the company's indifference and the FAA's failure in its duty of care, McDonnell Douglas had to deal with the biggest lawsuit in civil aviation history up to that date.

Three days after the death of those 346 people the FAA finally issued its directive to McDonnell Douglas and to all DC-10 operators. McDonnell Douglas continued to produce planes, but after a disastrous Sino-American merger and stiff competition from Boeing and Airbus, Boeing and McDonnell Douglas finally merged in a $13 billion stock-swap ($17 billion in today's terms) to form The Boeing Company in 1997.

9 September 1976

Flights: Inex Adria Aviopromet Flight 550 and British Airways Flight 475
Models: McDonnell Douglas DC-9-31 and Hawker Siddeley HS-121 Trident 3B
Registrations: UY-AJR and G-AWZT
Fatalities: 113/113 (108 passengers and 5 crew) and 63/63 (54 passengers and 9 crew), total 176
Principle cause: midair collision

BA 475 had left Heathrow at 08:32 bound for Istanbul. An hour and a half into its flight, it was cruising at 33 000 feet (10 000 metres) at a rate of 905 kilometres per hour (562 miles per hour) and approaching the ZAG VOR navigation beacon near Vrbovec, Croatia.

Inex 550 had left Split Airport in Croatia at 09:48 and was also approaching ZAG VOR on its way to Cologne in Germany. At 10:14 Inex 550 reported that it was climbing to 32 500 feet, but Zagreb ATC requested that they level off at their present altitude.

Barely two or three seconds later the outer part of Inex 550's left wing cut through the cockpit of British Airways 475. The wing sheared off and caused an explosive decompression, disintegrating British Airways 475's front fuselage. What was left of the plane crashed tail first into the ground below at about the same time that the crippled Inex 550 struck the ground tail first.

A TRAGIC MILESTONE

The collision remains the worst air accident in Croatia's history.

The investigation determined that, 'Improper ATC operation … Non-compliance with regulations on continuous listening to the appropriate radio frequency of ATC and non-performance of look-out duty from the cockpits of either aircraft,' were the ultimate causes of the accident.

The entire shift of controllers at Zagreb ATC was arrested. One controller was found guilty of criminal negligence and served two years of a seven-year sentence.

25 September 1978

Flights: Pacific Southwest Airlines Flight 182 and private plane
Models: Boeing 727-214 and Cessna 172
Registration: N533PS and unknown
Fatalities: 135/135 (128 passengers and 7 crew), 2/2 (pilot and instructor) and 7 on the ground, total 144
Principle cause: midair collision

PSA 182 had come from Los Angeles International Airport. The late autumn weather was perfect and the PSA crew was making a visual approach to their final destination, San Diego.

The single-engined Gibbs Flight Centre Cessna had taken off from Montgomery Field at 08:16, having spent the morning practising instrument landings on short trips. At 09:00 the crew of the 727 was aware of the climbing Cessna and keeping track of it. Meanwhile, ATC instructed David Boswell, the student pilot of the Cessna and his instructor, Martin Kazy, to maintain VFR at or below 3500 feet (1000 metres). PSA 182 was at 2600 feet (790 metres) with its landing gear down, ready to land.

At 09:00:38 the Boeing received the following message: 'PSA 182. Lindbergh tower, ah, traffic twelve o'clock one mile a Cessna', but by 09:01:11 the crew of the 727 had lost sight of the Cessna. At 09:01:28 San Diego ATC's alert warning system computers warned controllers of a conflict, but since San Diego is a busy airport, and such alerts happened all the time, the controllers ignored it.

At 09:01:47 the smaller plane hit the 727 in the right wing, disintegrating instantly.

Los Angeles is only a short distance from San Diego, and the 727 had almost full fuel tanks. As the wing erupted into flames, observers on the ground below could even feel the heat. One witness even reported apples and oranges baking on trees. The aeroplane banked at 50 degrees to the right and hit the ground nose down at 09:02:04.

In today's crowded skies it is essential that pilots, ground control, technology and luck all work hand-in-hand to keep planes apart. Once two aircraft share the same air space things go from bad to worse very quickly.

Faced with the sheer scale of an air disaster even professionals often stop to just stare, trying to make sense of it all. This unfortunate Spantax DC-10 crashed at Malaga Airport during take-off in 1982. Miraculously only 46 of the 393 passengers and crew died, although many more were no doubt seriously injured.

FAMOUS CASUALTIES

The Soviet Pahtakor football team was on CCCP-65735 and its members died along with all of the other occupants of both planes.

The crash destroyed 22 houses in North Park, San Diego. Wreckage killed seven on the ground.

One woman had a body crash through her window, covering her in gore and blood. It was a mere shadow of what was going on outside as plane wreckage and body parts showered the suburb.

Among other findings, the investigation determined that David Boswell had failed to maintain his Cessna on its assigned heading and that he had failed to inform the tower. Had he done so, the collision would not have happened.

11 August 1979

Flights: Aeroflot Flight and Aeroflot Flight
Models: Tupolev TU-134A and Tupolev TU-134AK
Registrations: CCCP-65816 and CCCP-65735
Fatalities: 94/94 (88 passengers and 6 crew) and 84/84 (77 passengers and 7 crew), total 178
Principle cause: midair collision

CCCP-65816 was cruising to Kishinev, Moldova, at an altitude of 27 500 feet (8400 metres) over Dneprodzerzhinsk in the Ukraine.

This route intersected the flight path of CCCP-65735 en route from Donetsk to Minsk. The air traffic controller monitoring the planes in the area noted that they were on a collision course and he ordered CCCP-65735 to climb to 29 500 feet (9000 metres). ATC heard a muffled reply. The controller assumed that it was CCCP-65735 confirming the altitude change. Unfortunately the message was from another flight altogether. CCCP-65735 never heard the instruction.

Both planes were enshrouded in clouds so thick that they couldn't see each other and they collided at 10:35.

19 February 1985

Flight: Iberia Airlines Flight 610
Model: Boeing 727-256
Registration: EC-DDU, *Alhabra de Granada*
Fatalities: 148/148 (141 passengers and 7 crew)
Principle cause: pilot error

Flight 610 was routine and ordinary for most of its journey from Madrid to Bilbao, Spain. At 09:22 the crew commenced a standard approach at an altitude of 7000 feet (2100 metres).

Airports all over the world come in all shapes, sizes and, importantly, altitudes. As a result each airport has a minimum sector altitude that the flight crew program into their navigation systems so that they know where their plane is, not only in terms of their absolute height above sea level, but also their relative height to the ground.

Although Bilbao Airport is only 138 feet (42 metres) above sea level, the minimum sector altitude for Bilbao is 4354 feet (1327 metres) in order to account for surrounding topography.

At an altitude of 5000 feet (1524 metres), the crew of Flight 610 programmed the altitude alert system (ASS) for 4300 feet (1310 metres). The AAS was now primed to give ground proximity warnings (GPW) in accordance with the optimal path for a landing. In approach mode the warning sounds 900 feet (300 metres) prior to reaching the preset altitude, and, if the plane falls 300 feet (100 metres) below the programmed height, the GPW would sound again as a deviation warning.

This is where things went wrong. The crew of Flight 610 had programmed the AAS for 4300 feet when they were at 5000 feet. They were already at a point where they were 200 feet below the first warning. The GPW wouldn't sound now until they were at around

4000 feet (1220 metres). Sure enough, the GPW was working perfectly and sounded the warning at 4040 feet (1230 metres).

The problem was that the flight crew thought that this was only the first warning and that they were at 5200 feet (16 000). Apparently they didn't look at their altimeter to double check. The CVR even records the captain telling the GWP to 'Shut Up!' because he thought he was still in approach mode when he was actually 1200 feet (370 metres) lower than he thought he was.

At 09:27, 57 seconds after passing the minimum sector altitude, the 727 struck the base of a group of antennas on top of Mount Oiz. The plane was still 30 kilometres (19 miles) south east of Bilbao. The left wing broke off and the rest of the plane crashed into the lower slopes of the forested mountain.

8 January 1989

Flight: British Midlands Airways Flight 092
Model: Boeing 737-4YO
Registration: G-OBME
Fatalities: 47/125 (47 out of the 118 passengers and none of the 7 crew)
Principle cause: metal fatigue and pilot error

It was only weeks after the Lockerbie crash and BMA 092 from London to Belfast had been in the air for thirteen minutes when the crew heard a bang, and the smell of overheated metal permeated the cockpit.

They suspected engine trouble and they were correct. Metal fatigue had caused a breakage in a fan blade in the left engine. Metal fragments that had entered the engine were now damaging the interior mechanisms. The damage caused a compressor stall which began a violent shaking in the aircraft. The plane was only two months old.

The flight crew referred to the manufacturer's checklist, but didn't think things through properly. Two minutes after the vibrations had started, the pilots switched off the autopilot and powered down the right engine. The vibrations became less violent and the noise quietened down considerably, so the crew assumed that the problem was with the right engine. They ignored a small vibration indicator lamp for the left engine, which was telling them that it was near the maximum vibration; because the switch was both unalarmed and obscurely positioned, no one thought to look at it.

The captain then announced over the address system that they had shut down the right engine. This confused the passengers and three of the cabin crew, as they could still see sparks and flames coming out of the left engine, but no one spoke up, probably trusting in the experience and authority of the flight crew. The plane was among clouds, and the flight crew couldn't see the engines from the cockpit, but at no time did any of them leave the cockpit to look at the engines and check their assumptions against what was going on in the real world.

'... at no time did any of them leave the cockpit to look at the engines and check their assumptions against what was going on in the real world.'

It is difficult to reconcile the power of machines that can carry hundreds of people at great speed over vast distances with their fragility.

The pilots requested a landing at East Midlands Airport. The plane was designed to fly on one engine if necessary and the flight crew was relying on their 'good' left engine to fly them there. They did not declare an emergency.

Meanwhile the left engine was taking the full load of the flight. It was gradually destroying itself, but maintaining enough power so that, even as the 737 began its descent, the engine kept the plane in the air, lulling the crew even further into their false sense of security.

The runway was 3 kilometres (2 miles) miles away and the plane was only 900 feet (300 metres) in the air when the flight crew increased power to the left engine before its final approach. It was more than the engine could bear and it disintegrated so quickly that the fire alarm didn't even go off.

Flight 092 was now completely without power. As the plane's airspeed fell below 230 kilometres per hour (142 miles per hour), it dropped out of the sky. At 20:24 the jet hit the top of a tree line before hitting the ground, belly up, in a field. It then bounced up and skirted the M1 motorway, before slamming into the road's west embankment only seconds away from the end of the runway at East Midlands Airport. Miraculously, although the M1 is one of the busiest roads in the UK, there was no traffic in the vicinity when the plane came down.

Fortunately rescue services were on alert and reached the crash site within minutes. They were able to deal rapidly with the leaking fuel so there was no fire.

Some passengers, many understandably in a state of shock, were able to walk away from the wreck with only minor injuries. Others were trapped in the remains of the fuselage for hours. Rescuers removed the last of the dead at around 04:00 the next morning.

22 December 1992

Flights: Libyan Arab Airlines Flight 1103 and Libyan Air Force
Models: Boeing 727-2L5 and MiG-23UB
Registrations: 5A-DIA and unknown
Fatalities: 157/157 (147 passengers and 10 crew) and 2/2 (crew), total 159
Principle cause: midair collision

A TRAGIC MILESTONE

The Flight 1103 accident remains the worst air disaster in Libya's history.

Flight 1103 was on its final approach on Visual Flight Rules to Tripoli Airport. It had descended to 3500 feet (1066 metres) and was just 9 kilometres (6 miles) east of Tripoli when it collided with the MiG fighter.

23 March 1994

Flight: Aeroflot Flight 593
Model: Airbus A.310-304
Registration: F-OGQS
Fatalities: 75/75 (63 passengers and 12 crew)
Principle cause: pilot error

In the annals of aviation accidents there's nothing quite like Flight 593. Captain Yaroslav Kudrinsky and Second Pilot Piskarev were flying between Moscow and Hong Kong. At around 00:40 off-duty Aeroflot pilot Makarov decided to let Captain Kudrinsky's children have a turn sitting in the left pilot's chair. The captain's 11-year-old daughter, Yana, was happy to sit and watch while her father demonstrated some autopilot features. Then it was 15-year-old El'dar's turn. He asked for and received permission to turn the control wheel, which he did. In doing so he carefully returned the wheel to its neutral position. The captain then repeated the same demonstration to his son that he'd

shown to his daughter only a minute before.

Due to a design anomaly, the instrument changes had inadvertently disconnected the autopilot linkages to the wing ailerons, but kept the autopilot on. Although an indicator light flashed to signal that the ailerons were now under manual control, the crew didn't notice anything. Their previous experience on other aircraft had conditioned them to having an audible signal.

With no active control on the ailerons, the Airbus began to bank sharply to the right at a rate of 2.5 degrees per second. Ironically it was Kudrinsky's son El'dar who first noticed this. The instruments signalled that the plane was now in a holding pattern. This only confused the crew.

FROM THE COCKPIT

El'dar Kudrinsky: May I turn this, the control wheel, a little bit? ... Why is it turning?
Captain Kudrinsky: Is it turning by itself?
El'dar: Yes it is!
...
Captain Kudrinsky: The other way!
Co-Pilot Piskarev: The other way!
Captain Kudrinsky: Turn to the left!
Co-Pilot Piskarev: Back!
Captain Kudrinsky: To the left!!!!

A high altitude accident rains aeroplane debris over a wide area. Fortunately for those below, most accidents have occured over sparsely populated regions.

'It was standard procedure at this time to cover the sensors of 757s with duct tape during the wash. It was also standard procedure to remove the duct tape after a wash.'

The nose suddenly dropped sharply and the autopilot wasn't able to maintain altitude. When the banking reached 45 degrees, the plane started to buffet. Only now did the pilot ask the co-pilot to take control.

As the pilot struggled to regain his seat, the co-pilot tried to reach the control wheel, but, as his seat was fully retracted, it took him an extra two or three seconds to reach the controls. In those few seconds the plane banked to 90 degrees and pitched up so fast that its occupants experienced acceleration of 4.8 gs—in effect, they weighed almost five times their normal weight. The plane stalled and entered a spin. Over the next 2 minutes the co-pilot struggled through a series of stalls and rapid pull-ups. Although he almost managed to level the plane out and reach a stable flight configuration, the plane had lost so much altitude by this point that there was no room left to manoeuvre and the Airbus crashed in a remote forested and mountainous area of Russia a few hundred kilometres north of the border between Kazakhstan, Mongolia and China.

What the pilots were not aware of was that there was a recovery procedure for the 310. If they'd let go of the controls after having achieved a near vertical stall, the plane would have recovered on its own, automatically.

2 October 1996

Flight: AeroPeru Flight 603
Model: Boeing 757-23A
Registration: N52AW
Fatalities: 70/70 (61 passengers and 9 crew)
Principle cause: negligent homicide

Aeroplanes, just like cars, need a wash every now and then, but some aeroplane parts, like static air intake sensors, which determine airspeed and altitude, don't take kindly to having soapy water drip into them. It was standard procedure at this time to cover the sensors of 757s with duct tape during the wash. It was also standard procedure to remove the duct tape after a wash.

Flight 603 left Lima, Peru, for Santiago, Chile, at 00:42. As the plane climbed the captain noticed that his airspeed and altitude readings were too high and there was an overspeed warning. Simultaneously the co-pilot's airspeed warnings were too low. As usually happens in cases of low airspeed, the plane's systems activated a motor that shook the airstick to physically prompt the pilot to take action. What the flight crew didn't know was that duct tape was still covering the intake sensor.

When the plane started to behave erratically Pilot Eric Schreiber and Co-Pilot David Fernandez couldn't make sense of what the instruments were telling them. The plane was, in effect, lying to them about the real world. Constant alarms in the cockpit were not only distracting them, they had nothing to do with what the plane was actually doing.

The transcript clearly shows that in the ever-increasing confusion the pilot thought that they had control problems, when in fact the plane was responding perfectly to the pilots' control. Only later did they realise that the information they were receiving wasn't worth anything. Cutting engine power to stop the fictitious overspeeding only led to a constant danger of stalling. Other errors, like attempting to engage the autopilot and failing to make use of the radio altimeters for information, only made the crisis worse.

It was the dead of night. There was no way to reckon visually what was going on, so the crew had to rely on instruments that could not give them real information. Schreiber and Fernandez struggled for half an hour to make the instruments tell them that everything was alright, but the instruments never did.

The plane made its first impact at a speed of 480 kilometres per hour (300 miles per hour) with the left wing and engine down. It then bounced off the water, climbed 200 feet (60 metres) and flipped over. At 01:11, when the plane finally crashed into the Pacific, 73 kilometres (45.6 miles) north-west of Lima, the altimeter reading was 9700 feet (3000 metres).

Although the crew had failed to notice the taped-over sensors during the pre-flight check and the maintenance supervisors had some responsibility for implementing poor procedures, in the end the authorities decided to try only the maintenance worker for negligent homicide. Found guilty, he spent two years in jail.

AeroPeru was already dealing with considerable financial and management problems at the time of the crash and the airline went under on 10 March 1999. The families of the victims of Flight 603 received compensation from the liquidators of AeroPeru and also from Boeing on account of the 757's defective design.

12 November 1996

Flights: Saudi Arabian Airlines Flight 763 and Kazakhstan Airlines Flight 1907
Models: Boeing 747-168B and Ilyushin 76TD
Registrations: HZ-AIH and UN-76435
Fatalities: 312/312 (289 passengers and 23 crew) and 37/37 (27 passengers and 10 crew), total 349
Principle cause: midair collision

Flight 1907 was on its descent into Delhi, 120 kilometres (74 miles) from the Delhi Airport Beacon, when Delhi ATC cleared the flight to descend to and report at 15 000 feet (4500 metres).

Flight 763 had left Delhi at 18:32 bound for Dharan in eastern Nepal and was using the same air corridor as Flight 1907. Delhi ATC took this into account and asked Flight 763 to maintain 14 000 feet (4300 metres). Shortly afterwards Flight 1907 reported that it was now 75 kilometres (46 miles) away at 15 000 feet. The controller responded 'Roger. Maintain 150. Identified traffic 12 o'clock, reciprocal Saudia Boeing 747, 14 miles. Report in sight.'

The Kazakh crew queried the location of the oncoming aircraft. Delhi ATC replied 'Fourteen miles now, roger 1907.' In the silence that followed Delhi ATC reconfirmed. 'Traffic in 13 miles, level 140.' At that point both planes should have been 20 kilometres (12 miles) apart with the Kazakh plane 1000 feet (300 metres) above the Saudi Arabian 747.

Emergency air chutes are seldom used and seldom tested. Every time they are engaged the aircraft has to be taken back to the manufacturer for reinstallation and priming, at a cost of tens of thousands of dollars.

The Concorde has only ever suffered one major hull loss—a freak accident involving a small piece of metal, a torn tyre and a shockwave that ignited a fuel tank. The result—113 dead and, ultimately, the withdrawal of the supersonic jet from service.

Unknown to ATC and the Saudis, the Ilyushin had descended to 14 500 feet then to 14 190 feet (4300 metres). The equipment at Delhi ATC was outdated and did not give the controllers accurate height information as well as horizontal separation. The Indian Government later insisted that the blame fell entirely on the Kazakh aircrew, but the explanation rang hollow. In the six months bridging 1994 and 1995, India had already had three near air collisions

The Saudi Arabian plane had only been airborne for seven or eight minutes when, at 18:40, both aircraft collided into each other, instantly becoming fireballs, which plummeted to the ground. There is evidence that the Saudi captain was alive long enough and had enough control to guide his 747 away from the village of Charki Dadri. The plane crashed in a farming district 5 kilometres (3 miles) from the town and left a trench 60 metres (180 feet) long and 5 metres (15 feet) deep. When the villagers came to the site to see what they could do to help, they were too shocked to react.

The Ilyushin 76TD crashed 11 kilometres (7 miles) away with wreckage and body fragments strewn the whole way between the planes. In the absence of nearby morgue facilities, authorities brought the remains back to the airport and stored them on blocks of ice in the hangars.

A TRAGIC MILESTONE

At the time of writing this 1996 accident outside Delhi remains the worst midair collision in aviation history.

25 July 2000

Flight: Air France Flight 4590
Model: Aérospacial / BAC Concorde 101
Registration: F-BTSC
Fatalities: 109/109 (100 passengers and 9 crew)
and 4 on the ground, total 113
Principle cause: runway debris

At 14:37 a Continental Airlines DC-10 departed Charles de Gaulle Airport. Unknown to anyone at the time 45-centimetre (18-inch) strip of titanium fell off one of its engines and was lying on the runway.

AF 4590 bound for JFK-New York was supposed to have left over an hour earlier, but had been delayed. The flight crew had requested a replacement of the number two engine pneumatic thrust reverser motor as well as other parts. Due to another oversight the plane was one ton over the maximum take-off weight and fully loaded with 95 tons of fuel.

The Concorde began its take-off at 14:42:31. At 14:43:04 a front tyre of the left main landing gear ran over the metal strip and was torn apart so violently that a 5 kilogram (11 pound) slab of rubber detached and hit a fuel collector tank on the port wing. The shockwaves from the impact concentrated on the full tank, causing the tank wall to rupture from the inside. Fuel began to leak out, and aerodynamic forces fed the fuel into the left engine intakes.

The fuel caught fire and spread under the left wing. The left engines lost thrust. The flames set off a false alarm of a fire in number two engine. Pilots Jean Marcot and Christian Mary shut it down, as flames trailed 200 feet behind the plane.

Still the plane was climbing, and the pilots were doing their best to keep the nose up. Airspeed was now 370 kilometres per hour (230 miles per hour) and the attempt to retract the landing gear in order to decrease drag and increase airspeed failed.

The plane was now trying to reach nearby Le Bourget Aerodrome and had been in the sky for only two minutes when it rolled left. Engines three and four on the right wing lost airflow and stopped working. The Concorde crashed into the Hotelissimo Hotel at 16:44. All onboard the plane were killed as well as four people in the building.

On 16 August 2000 authorities revoked the Concordes' airworthiness certificate and they grounded all Concordes pending major changes to the fuel tanks. Concordes did not fly again until 7 November 2001. This was the first, worst and ultimately the only loss of a Concorde in aviation history.

8 October 2001

Flights: Scandinavian Airlines (SAS) Flight 686 and private plane
Models: McDonnell Douglas MD-87 and Cessna 525A Citation Jet 2
Registrations: SE-DMA and D-IEVX
Fatalities: 110/110 (104 passengers and 6 crew), 4/4 (2 passengers and 2 crew) and 4 on the ground, total 118
Principle cause: midair collision

It was 08:10 in the morning at Milano-Linate Airport, Italy. A heavy fog enshrouded the field, reducing runway visual range (RVR) to just 225 metres (740 feet) and it was growing worse all the time. Traffic volume at the airport was high. Radio communications between aircraft and planes were being conducted in both English and Italian, controllers were not conforming to international standards of phrasing and the aerodrome markings and signage did not comply with international standard regulations.

A new Cessna business jet was on the ground awaiting take-off. Authorities had not properly checked the flight crew's qualifications, and the enquiry later determined that authorities also failed to brief the Cessna crew with the correct information for negotiating the airport.

The Cessna was carrying one of the company's sales reps and a prospective customer, but the fog was in the way of a good demonstration that could lead to a sale. The small jet was parked in the airport's west apron pending clearance to leave. The occupants were in a hurry.

In short, it was an accident waiting to happen.

Meanwhile SAS 686 bound for Copenhagen was parked in the north apron. It had originally been scheduled to leave at 07:35, but it didn't receive engine-start clearance until 07:54. ATC then instructed SAS 686 to taxi to runway 36R.

Several minutes later ATC gave the Cessna pilot start-up clearance and instructed the plane to enter taxiway R5. The Cessna followed the yellow taxi guideline but erroneously entered taxiway R6.

It was now 08:09 and ATC cleared SAS 686 for take-off. At the same time the Cessna had reported its position as 'S4'. ATC cleared the Cessna to continue its taxi even though position 'S4' was meaningless and in fact ATC didn't know where the Cessna actually was.

While SAS 686 was speeding down runway 36 for take-off, the Cessna crossed the runway holding sign, the equivalent of running through a red light at an intersection. At 08:10:21 the pilots of SAS 686 must have seen the Cessna on the runway to their right and immediately pulled the nose up. The nose landing gear was already in the air, but the main landing gear was still on the ground when the passenger jet's right wing hit the Cessna at 270 kilometres per hour (167 miles per hour) and instantly killed the small plane's occupants.

Rescue workers remove the remains of the wreckage of the crashed Scandinavian Airlines System (SAS) MD-87 aeroplane at Linate Airport, near Milan, 8 October 2001. The SAS jet crashed into an airport hangar after colliding with a small plane in heavy fog, killing 118 people including four workers in the hangar. This was the worst aviation accident to happen in Italy in the last 30 years.

Families of the victims of Bashkirian Airlines Flight 2937 had considerable access to the crash zone and were even involved in the recovery. Here they gather, the engine of the downed Tupolev serving as a grave marker— a huge headstone of metal —a memorial of tortured machinery.

The MD-87's wing was damaged and the right landing gear collapsed and further damaged the number two engine, which then fell off its mounting. The plane reached a height of just 35 feet (11 metres) and been airborne for only twelve seconds before crash debris entered the left engine. The now-damaged left engine couldn't provide enough power and the plane landed heavily on its left gear with the stump of the right gear and the right wingtip gouging the runway. The pilot applied maximum reverse thrust from his remaining engine, but the plane skidded out of control, onto grass, across a service road and into a baggage handling building near the end of the runway. Four baggage handlers were inside. With two wingloads full of fuel, the plane crumpled. No one survived the ensuing fire as the building collapsed over the wreck.

In 2004 a Milan court found Airport Director Vincenzo Fusco and an air traffic controller guilty of negligence and they were sentenced to eight years in prison. Six others received lighter sentences. The Italian Parliament issued a pardon on 29 July 2006, effectively reducing the prison terms by three years.

There is now a memorial park near the airport, the Bosco dei Faggi, which contains 118 beech trees to represent the crash victims.

1 July 2002

Flights: Bashkirian Airlines Flight 2937 and DHL Aviation Cargo Flight 611
Models: Tupolev TU-154M and Boeing 757-23APF
Registrations: RA-85816 and A96-DHL
Fatalities: 69/69 (60 passengers and 9 crew) and 2/2 (crew), total 71
Principle cause: midair collision

Although not one of the worst accidents in terms of loss of life, the story of Flight 2937 has its own unique tragedy that extended far beyond the crash itself.

It began in the town of Ufa in western Russian. UNESCO selected 46 of the city's most talented and gifted children to go on a two-week vacation to Barcelona. The tourist agency that had arranged their itinerary made a mistake and took the children to the wrong airport. The group missed their flight and had to spend two days in Moscow waiting for a charter flight to take them to Spain. The group finally left on Flight 2937 shortly before 23:00. The nominal pilots of the plane were Alexander Gross, with over 30 years experience, and First Officer Oleg Gregoriev, the airline's Chief Pilot who was undertaking an evaluation. Three other crew filled the cockpit. There was no end of expertise on the plane.

Among the few people on the flight who were not part of the school group were the Kaloyev family—a mother and two children who were going to meet their father, Vitaly, in Barcelona, where he was working as an architect on a construction project.

The evening run would take them over southern Germany. DHL 611 was a cargo run from Bergamo, Italy, to Brussels, Belgium, with no passengers. Captain Paul Philips and First Officer Brant Campioni were in the cockpit and took off at 23:06.

Both planes were fitted with a traffic collision avoidance system (TCAS), which continually monitored the plane's position, heading, speed and height.

The DHL evening run would also take the plane over southern Germany. During the course of a long flight, air traffic control guidance passes from one centre to another depending on the position of the aircraft. A private company, Skyguide, monitors and controls the airspace over northern Switzerland and southern Germany. There were only two controllers on duty that night, but, because of the light traffic, one took an extensive break, leaving controller Peter Nielsen on his own to handle all incoming information from two workstations. At 23:10 maintenance staff informed Nielsen that they had work to do on the radar. This meant that the monitoring systems would work more slowly—a problem that would become critical later. Even worse, the technicians had to shut down the telephones, leaving the control room with only stand-by phones.

Anticipating possible problems Nielsen tried to find back-up from Friedrichshafen Control in Germany several times between 23:25 and 23:33. His attempts failed because the stand-by phones weren't working.

At 23:30:11 control of Flight 2937 went from Munich to Skyguide in Zurich, and an entire region of airspace was now in the hands of only one controller using compromised equipment. Events came to a head when several things happened within a very short period of time. Nielsen granted a request from DHL 611 to climb to 36 000 feet (11 000 metres)—the same level as Flight 2937. Then another plane, Aerolloyd 1135 needed his attention. Almost immediately Flight 2937 needed him, then northbound Thai Airways Flight 933 needed him too. In the constant demand from four different planes with a system that was running sub-optimally, Nielsen wasn't aware that DHL 611 and Flight 2937 were on a collision course travelling at a relative velocity of over 1300 kilometres per hour (807 miles per hour).

The short term conflict alert system (STCA) may not have been working at Skyguide, but the STCA at Karlsruhe Upper Area Control Centre (UAC) in Germany was. International rules designed to avoid confusion prevented the controller from contacting either the Russians or the cargo plane directly, and between 23:33 and 23:35 the controller tried to contact Skyguide, but couldn't get through because the phones were down. Even the priority button failed.

At 23:34:42 the Russians' TCAS gave them a warning. Seconds later the DHL TCAS gave its warning. Both machines on both aircraft were working perfectly. The DHL plane was below the Russians and a mechanical voice began to chant 'Descend. Descend' incessantly. The Russian's TCAS advised the crew to 'Climb. Climb.'

At this point the lagging screens told Nielsen that there was the possibility of a collision, and he made a critically bad call. He told Russian Flight 2937 to: 'Descend flight level 350. Expedite. I have crossing traffic'. This contradicted the Russian plane's TCAS system. When Nielsen repeated the instruction, both planes followed his advice—the DHL because the controller was reinforcing the message from the TCAS and the Russians because in a critical situation they believed that a human being had more authority than a machine. This decision was backed up by the plane's Flight Operations Manual, but contradicted the TCAS Flight Operations Manual.

Eight seconds before the collision, both planes made visual contact and realised that they were actually diving towards each other. The crews did what they could; it wasn't enough.

At 23:35:32 the aircraft collided. The tail fin of the 757 struck the left of the Tupolev fuselage. The collision sheared off nearly all 757's tail fin and the whole front of the Tupolev fuselage. A second later the left and right wings of the Tupolev broke off from the rear fuselage and tail assembly. DHL 611 struggled on for two minutes while losing engines from the shock of the impact and the aerodynamic forces of the fall. The remains of the plane landed 7 kilometres from the impact point.

Eyewitnesses reported orange flaming fragments of the planes lighting up the overcast sky and the sound of continuous explosions. The fragments of the Tupolev

Forty-six children died on Bashkirian 2937—all from the town of Ufa in Western Russia.

Over 6000 people converged on the area around Uberlingen, Germany, in July 2002 to face sights like this—devastation in the midst of beautiful woodland. This is part of DHL Cargo Flight 611.

'The fragments of the Tupolev and the bodies of the children who had been onboard rained down over the town of Uberlingen in Germany.'

and the bodies of the children who had been onboard rained down over the town of Uberlingen in Germany. Within moments of the crash, Peter Nielsen's colleague returned from his break and Nielsen had to be led out of the room. Distraught and traumatised he never returned to the control room.

The days of controllers taking long breaks while on duty were over. Skyguide reduced operations for the next three weeks and the tragedy marked the beginning of a sweep of reforms.

Six thousand people were involved in the search over the next week. One of them was Vitaly Kaloyev who had flown in from Barcelona. He found the body of his four-year-old daughter, miraculously intact and lying in a wood. Days later, other searchers found the mutilated bodies of his wife and son.

In the aftermath of the accident, several further incidents drove Vitaly over the edge. He spent a year practically living at the graves of his family in Ufa. On the first anniversary of the crash, at a ceremony in Uberlingen, Vitaly asked the director of Skyguide, Alan Rossier, if he could meet the controller who had been on duty on the night. The executive ignored the question. Vitaly later received a letter from a law firm in Hamburg offering him 160 000 Swiss Francs ($130 000) for the loss of his family—if he declined any further action against Skyguide. This just drove him from despair to anger.

On 24 February 2004 Vitaly tracked down Peter Nielsen at his home in Zurich. Vitaly maintained that he wanted Peter Nielsen to apologise for the deaths of the Kaloyevs, but according to Vitaly, Peter Nielsen then,

'hit me on the hand when I was holding the envelope with the photographs of my children. I only remember that I had a very disturbing feeling, as if the bodies of my children were turning over in their graves.'

Vitaly claimed that he did not remember what happened afterwards. He stabbed Nielsen to death. A few days later the police found him in a hotel room in a state of shock.

He was finally sentenced to eight years in prison. As of this writing he's still in jail in Zurich with a release date pending in 2011.

Legal action against Skyguide continues. Although Skyguide has accepted responsibility for the accident and has paid out compensation to the families, the court of Konstanz determined that the German government was at fault, as Skyguide had no legal right to offer air traffic control services in German airspace. On 7 August 2006 prosecutors in Winterthur, Switzerland, filed 'homicide by negligence' charges against eight Skyguide employees.

To this day members of Skyguide staff keep a white rose at Peter Nielsen's former workstation.

29 September 2006

Flights: Gol Transportes Aereos Flight 1907 and
ExcelAire Delivery Flight
Models: Boeing 737–8EH and Embraer Emb-135BJ
Legacy 600
Registrations: PR-GTD and N600XL
Fatalities: 154/154 (148 passengers and 6 crew)
and 0/7
Principle cause: midair collision

At 15:35 Flight 1907 departed Manaus, north Brazil,
and was expected to arrive at the Brazilian capital of
Brasilia at 18:12.

At the same time the brand new Legacy 600
Executive Jet was on its delivery run from its factory
at Sao Jose Dos Campos, to Ronkonkoma in New
York State. After leaving Brazilian airspace at an
altitude of 37 000 feet (11 300 metres) the Legacy
should have descended to 36 000 feet (11 000).
However, flight recorder transcripts leaked to the
press have revealed that ATC Brasilia told the Legacy
to maintain a level of 370 and that another ATC
reinforced the instruction to maintain that altitude.

At 16:56 the Legacy and the 737-800 collided over
the Amazon Rainforest. The 737 gradually lost control
and crashed into the forest, eleven minutes later
at 16:59. It was the first ever loss of a 737-800 and is
currently the worst aviation disaster in Brazilian history.

It was fourteen hours before rescuers found the
wreckage of the 737, 200 kilometres (120 miles) east
of Peixoto de Azevedo in the Jarina farm area. The
geography is mountainous and densely wooded.

Some 200 people were involved in the recovery
operation, which included five planes and three
helicopters from the Brazilian Airforce and even a group
of local Kayapo Indians. Although investigators found
the flight data recorder on 2 October it took almost
four weeks of intensive searching to find the cockpit
voice recorder. The field investigation was called off on
16 November 2006 after rescuers had recovered the
remains of 153 of the 154 victims.

The Legacy sustained only relatively minor damage
and managed to make an emergency landing at a
nearby Air Base at Serra de Cachimbo. Preliminary
investigations have revealed that the Legacy's
transponder was either switched off or inoperative due
to equipment failure at the time of the collision. Certainly
the jet's proximity alert system wasn't working either.

For the narrative of what happened to the survivors
of the Embraer Legacy and the aftermath of the
investigation to date see page 264.

Some terrain is more difficult than others. You have to reach the jungles of the more remote regions of Brazil by helicopter—if you are to reach them at all in the crucial hours after an air disaster when some of the seriously injured survivors might still have a chance of making it out alive.

ACCORDING TO AMERICA'S NATIONAL Transportation Safety Board,
the chances of dying in a plane crash is 52.6 million to one—in a single trip.
Considering that the rich and famous spend a lot more time travelling on planes,
and they make more trips, it's no surprise that, proportionally, there are more
celebrity deaths in aircraft accidents than those of 'ordinary people'.

Unfortunately there's no clear definition of 'famous'. One person's celebrity is
another person's 'has-been' or 'non-entity', so it's difficult to put together accurate
statistics—you can't even make them up. Nevertheless the following pages might
give you even more reason to feel safe.

PART FIVE:
Famous casualties of aeroplane crashes

The overwhelming majority of 'celebrity deaths' in aircraft occur on chartered flights or when the celebrity is flying the plane himself. It's nearly always a male pilot, women celebs are generally smarter.

Since even such accomplished pilots as Kingsford-Smith and Amelia Earhart died in aeroplane crashes, perhaps it might be a good idea if people with a talent for being famous in other fields stay out of cockpits and leave the flying to the pros.

Charles Stewart Rolls

Roald Amundsen

Charles Stewart Rolls

12 July 1910

The 33-year-old co-founder of Rolls Royce died when he fell and cracked his skull when his Wright Flyer broke up in midair only 20 feet (6 metres) above the ground. This was the first ever air fatality in Britain.

Harriet Quimby

1 July 1912

The pioneer aviator was the first woman to die, aged 37, in an air crash when her Blériot Monoplane pitched forwards during an air show. Quimby and fellow pilot William Willard were hurled out of the plane. Quimby was the first woman in America to earn a pilot's licence and the first woman to cross the English Channel on 16 April 1912—a feat that would have grabbed the world's attention if the *Titanic* hadn't sunk two days earlier. Amelia Earhart later credited Quimby for her bravery and inspiration.

Roald Amundsen

22 June 1928

The great Norwegian explorer and leader of the first ever expedition to successfully reach the South Pole, died when his French Latham 47 Flying Boat crashed into the Barents Sea, in the Arctic circle. He was on route to Spitsbergen as part of an international team to rescue the crew of the crashed airship *Nobile*. Although rescuers saved nine of the sixteen members of *Nobile*'s crew, no one ever found the body of Amundsen or of the other five men onboard the seaplane.

Denys Finch Hatton

14 May 1931

The professional big game hunter died when his plane stalled and crashed shortly after take-off in Voi, Kenya. He would have been obscure had not Karen Blixen, writing as Isak Dinesen, immortalised him in her novel

Amelia Earhart

Carole Lombard

Out of Africa, which was made into an Academy Award winning film in 1985.

Sir Charles Kingsford Smith

8 November 1938

The most famous of Australian aviators and his crew were the first men to cross the Pacific from America to Australia in 1928. The 38-year-old national hero and his mechanic disappeared when his Lockheed Altair VH-USB, *Lady Southern Cross*, vanished off the southern coast of Burma, near Aye Island, on 8 November 1935. Although wreckage of the plane showed up on the south Burman coast eighteen months later, no remains were ever found. Sydney's Kingsford Smith Airport is named after him.

Amelia Earhart

2 or 3 July 1937

The 40-year-old pilot and her navigator Fred Noonan (44) disappeared off Lea, New Guinea. The subjects of the most extensive and expensive search operation in history up to that time—costing $4 million and covering 640 000 square kilometres (250 000 square miles) of ocean—neither the aviators nor their Lockheed 10E Electra were ever found.

Carole Lombard

16 January 1942

The 33-year-old actress and adored wife of Hollywood legend Clark Gable, was on a special tour promoting war bonds when the Douglas DC-3 that she was on inexplicably drifted off course. At 19:23 Transcontinental Flight 3 crashed into Mount Potosi near Las Vegas, Nevada, killing Lombard, her mother, and her press agent as well as nineteen other passengers and crew.

Leslie Howard (centre)

Leslie Howard

1 June 1943

The British matinee idol, famous for his role as Ashley Wilkes in *Gone with the Wind*, was returning from Lisbon on a British Overseas Airways Douglas DC-3 when the Luftwaffe shot the aeroplane down off the coast of France, killing all seventeen onboard. Unfortunately for the 42-year-old actor, his manager Alfred Chenhalls looked like the British Prime Minister Winston Churchill, while Howard himself was of similar build to Churchill's bodyguard, and one theory posits that the Germans shot down the plane as a result of mistaken identity.

Glenn Miller

15 December 1944

The 40-year-old big band leader and Air Force Major, as well as two other officers and the pilot of the single-engined Noorduyn Norseman he was travelling in, disappeared over the English Channel. One of the most common explanations is that the plane had flown into a safe drop zone—an area where returning bombers drop unused bombs on their way back from unsuccessful bombing runs—and that the plane was accidentally hit, a tragic example of friendly fire. Neither the wreckage of the plane nor the bodies were ever recovered, although in 1992, at the request of his daughter, Miller received a headstone in Arlington National Cemetery.

'... his manager ... looked like the British Prime Minister Winston Churchill ... and one theory posits that the Germans shot down the plane as a result of mistaken identity.'

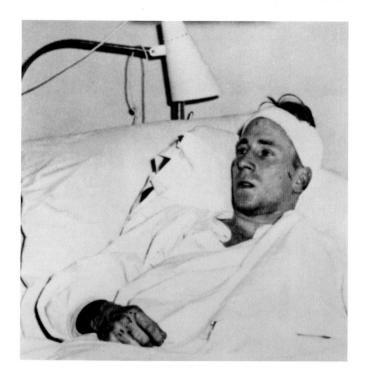

Manchester United Football Team survivor, Bobby Charlton.

Mike Todd

Manchester United Football Team

6 February 1958

Eight members of the British soccer team, with an average age of only 24, died when British European Airways Flight 609 crashed at the end of its runway at Munich Airport in Germany. The twin-engined Elizabethan was attempting take-off during a heavy snowstorm after two previous failed attempts. Later investigation concluded that the plane failed to achieve take-off speed because of the build-up of ice slush on the runway.

The plane overshot the runway, hit a house with its port wing, then veered right, hit another building and burst into flames. However, the plane's fuselage didn't catch fire and uninjured passengers were able to go back into the wreckage to rescue the injured. This accounts for the relatively high survival rate. Of the 44 onboard the flight, 23 in total died. Among the survivors was Bobby Charlton, who went on to win 109 caps for England, retiring in 1973 and becoming a director of Manchester United in 1984.

Mike Todd

22 March 1958

Although the film director and producer developed an excellent 70mm format for wide-screen film, the 49-year-old is often only remembered for being Elizabeth Taylor's third husband. He, along with screenwriter and biographer Art Cohen (50), were killed when the Lockheed Loadstar they were in crashed near Grants, New Mexico, at 02:05. Taylor was at home with bronchitis. Mike Todd, ironically, had named the plane the *Lucky Liz*.

Egyptian Fencing Team

14 August 1958

A few months after the Manchester United tragedy, KLM Flight 607E was flying over the North Atlantic, about 160 kilometres (100 miles) west of Ireland, when the Lockheed Constellation's mysteriously crashed killing all 99 onboard, including six members of the Egyptian fencing team. The plane crashed into deep water and was not recovered.

Buddy Holly, Ritchie Valens and 'the Big Bopper' Jiles Perry didn't walk away from this wreckage.

Buddy Holly and Ritchie Valens

3 February 1959

Possibly the most famous of all of the 'famous fatalities'. Inadequately qualified pilot Roger Peterson attempted a take-off from Mason City, Iowa, in a Beechcraft Bonanza during a snowstorm. The icing on the wings proved fatal for the pilot and his three passengers, Ritchie Valens (17), Jiles Perry 'The Big Bopper' Richardson (29) and Buddy Holly (22).

Magnanimity and fate played their roles here. Holly had chartered the plane to take him and the Crickets to North Dakota as part of a Winter Dance Party Tour. The Big Bopper had the flu, so Waylon Jennings generously gave up his seat for him. Valens flipped a coin and won the toss against Tommy Allsup, so both Jennings and Allsup had to take a bus. Ironically, this meant that when the small plane crashed at around 01:00 into a cornfield

the Crickets survived without Buddy Holly. The events of that morning were immortalised in the films *The Buddy Holly Story*, *La Bamba* and most significantly in Don McLean's ballad *American Pie*.

In his 1996 autobiography, Waylon Jennings recounted that Buddy Holly joked that he hoped that the bus that Jennings and Allsup were taking would stall. Jennings joked back that he hoped the plane would crash—a comment that did nothing to alleviate Jennings' survivor guilt.

American Figure Skating Team

15 February 1961

Sabena Flight 548 near Brussels, Belgium, nosedived and crashed when the plane's flight controls failed. 72 people died, including eighteen members of the American figure skating team.

Dag Hammarskjöld

Patsy Cline

Green Cross Chilean soccer team

3 April 1961

A Douglas DC-3 LAN flight 310 crashed into the Las Lastimas mountains near Llico, Chile. The cause remains officially unknown, but may have been due to icing on the wings. Result: all of the 24 passengers and crew killed, including the Green Cross Chilean soccer team. *Lastima* is Spanish for pity or lament.

Dag Hammarskjöld

17 September 1961

The United Nations Secretary General and fifteen others died when their Douglas DC-6B crashed into the jungle near Ndola, Zambia. Conspiracy theories aside, it is possible that the flight crew miscalculated the landing approach because they used altitude data for N'dolo, Congo, rather than Ndola (then northern Rhodesia).

Patsy Cline

5 March 1963

The iconic country and western singer, aged 30, died along with Hawkshaw Hawkins (39), Cowboy Copas (49) and Cline's manager, Randy Hughes, when the pilot of a Piper Comanche flew them into a storm over Camden, Tennessee County, at around 18:20.

Jim Reeves

31 July 1964

The country and western singer was piloting a Beechcraft 35-B33, 16 kilometres (10 miles) from Nashville, Tennessee, when he flew into a storm at 16:52. It took authorities two days to find the wreckage and the bodies of Reeves and his manager Dean Manuel.

Otis Redding

10 December 1967

The 26-year-old singer and four members of his Bar-Kays band were killed when the Beechcraft H18 they were flying in through fog crashed into Lake Montana at 15:25.

Mohammed bin Laden

29 May 1968

The 73-year-old father of the more infamous Osama bin Laden, died when the Beech 95-C55 he was in crashed on take-off. The pilot had the fuel selector in the wrong position and the plane simply ran out of fuel.

Rocky Marciano

31 August 1969

An inexperienced pilot handling a Cessna 172H Skyhawk crashed into a solitary oak tree while descending into Newton, Iowa, when bad weather forced him to attempt a landing. The heavyweight champion, the pilot, and the son of one of Marciano's friends died in the accident. The man born Rocco Francis Marchegiano was on his way to a surprise party for his 46th birthday.

The Strongest, the Bolivian soccer team

26 September 1969

For reasons unknown, a Lloyd Aereo Boliviano Douglas DC-6B crashed into a mountain near La Paz,

Rocky Marciano

Bolivia, at an altitude of 15 500 feet (4700 metres). All 74 passengers onboard were killed, including the entire Bolivian soccer team.

Audie Murphy

28 May 1971

All of the five onboard an Aero Commander 680 died when the plane crashed into the mountains near Roanoke, Virginia. Among them the World War II hero and film star Audie Murphy, aged 47.

Alexander Onassis

22 January 1973

The son of shipping billionaire Ari Onassis was only 24 when the Piaggio he was flying reached an altitude of 100 feet (30 metres) then rolled to the right and nosedived into the ground at Athens at 10:15.

Audie Murphy

Jim Croce

20 September 1973

All of the six people onboard a Beechcraft E18S died when it crashed into trees at the end of the runway at Natchitoches, Louisiana, at 22:45. The pilot, Robert Newton Elliot, had failed to gain sufficient height. The passengers comprised publicist Kenny Cortese, musician Maury Muehleisen (24), road manager Dennis Rast, comedian George Stevens and, most famously, singer Jim Croce (30).

Michael Findlay

16 May 1977

The schlock horror director of such endearing titles such as *The Ultimate Degenerate*, *Shriek of the Mutilated* and *Virgins in Heat*, met with an end eerily redolent with perverse irony. As he was boarding a helicopter on the roof of the Pan Am Building in New York the copter's landing gear collapsed from metal fatigue. As the rotor blades tilted they decapitated Findlay and killed three other passengers. A detached rotor blade also killed a woman on the street below.

Francis Gary Powers

1 August 1977

After becoming famous for piloting the U2 spy plane shot down by the Russians Francis Powers became a news reporter for Channel 4 in California. He and the pilot of the Bell 206B Jetranger helicopter died when it ran out of fuel and crashed in Encino after covering a fire story.

Lynyrd Skynyrd

20 October 1977

A malfunctioning magneto gave false fuel readings to pilot Walter McCreary, and co-pilot William Gray. When the chartered Convair CV-300 they were piloting ran out

> 'The schlock horror director ...
> met with an end eerily redolent
> with perverse irony.'

Lynyrd Skynyrd's Cassie Gaines (right) with
Leslie Hawkins (centre) and Jo Jo Billingsley (left).

of fuel near Gillsburg, Mississippi, they lost control and crashed in the woods below. Both died instantly. Of the 24 passengers onboard four others died. Three were members of the southern rock group, the Lynyrd Skynyrd Band: lead singer Ronnie van Zant (29), vocalist Cassie Gaines (29) and lead singer/guitarist Steve Gaines (28). Their assistant road manager Dean Kilpatrick also died.

To literally add insult to injury, drummer Artimus Pyle ran nearly a mile suffering from several broken ribs to the house of farmer Johnny Mote, who upon seeing the delirious blood-splattered 'hippy' decided that the best way to deal with him was to give him a gunshot to the shoulder. Pyle survived.

Shortly before the crash Lynyrd Skynyrd had released their sixth album, *Street Survivors*. The original cover showed the band members engulfed with flames. After the crash the record company MCA withdrew the album and re-released it with the band simply set against a black background. Needless to say, the original cover is now a collector's item.

Sanjay Ghandi

23 June 1980

The n'er-do-well son of Indira Ghandi and the elder brother of the more substantial Rajiv, Sanjay Ghandi (34) was piloting a Pitt Special stunt plane, with which he was unfamiliar and which had not been properly tested. He and his co-pilot died when it crashed in Delhi.

Boris Sagal

22 May 1981

The 'long and relatively undistinguished' career of Boris Sagal (53)—father of actress Katy Sagal and director of some episodes of television series such as *The Twilight Zone*, *Alfred Hitchcock Presents* and *Columbo* as well as the cult movie *The Omega Man*—ended in a way very similar to Michael Findlay. He accidentally walked into the tail rotor of a helicopter, which almost decapitated him. The helicopter was parked on the grounds of the Timberline Lodge on the south side of Mount Hood, in Oregon. The Timberline would later be used for the exterior shots in the film *The Shining*.

Vic Morrow

23 July 1982

When a Bell UH-1B helicopter crashed during the filming of the movie *The Twilight Zone* it received worldwide attention. Shrapnel from a special effects explosion

Vic Morrow

Ricky Nelson

hit a tail rotor, bringing down the copter. Actor Vic Morrow (57), the father of Jennifer Jason Leigh, was on the ground below when it crashed into him and two child actors—Myca Dinh Le (7) and Renee Shin-Yi Chen (6). The rotor blades decapitated Morrow and one of the children, while the other was crushed to death beneath the helicopter wreckage. Child labour laws and safety standards on movie sets in California underwent considerable revision and reform in the wake of the incident.

Ricky Nelson

31 December 1985

Singer and former teen idol Ricky Nelson (45), his fiancée Helen Blair, members of his back-up group The Stone Canyon Band—Andy Chapin, Rick Intveld, Bobby Neal and Patrick Woodward—as well as soundman Clark Russell, all died when the Douglas DC-3 they were on crashed into a field in De Kalb, Texas. The crash happened when smoke filled the cabin and cockpit of the plane,

probably the result of a defective cabin heater. The pilots lost control of the plane but survived the crash itself.

Dean Paul 'Dino' Martin

21 March 1987

Dean Martin's 35-year-old son was doing National Guard service in his F-4 phantom fighter jet. While performing a maximum take-off climb, the plane entered clouds and crashed, upside down, into the granite face of Mount San Georgio, California. He and Radio Intercept Officer Ramon Ortiz died in the crash.

Salem bin Laden

29 May 1988

Continuing the family tradition of unhappy incidents involving aeroplanes the 42-year-old brother of Osama died when the Ultralight he was piloting crashed into power lines near San Antonio, Texas.

The coffin of former Burundi President Cyprien Ntaryamira.

An F-14 Tomcat as flown by Lieutenant Kara Spears Hultgreen.

Stevie Ray Vaughn

27 August 1990

Pilot error at 01:00 in a Bell BHT-206B was responsible for the deaths of all five onboard when the helicopter failed to gain sufficient altitude to clear rising terrain in fog in Elkhorn, Wisconsin. The 35-year-old legendary blues guitarist was among the dead.

Zambian National Soccer Team

27 April 1993

Eighteen members of the Zambian National Soccer Team were among the full compliment of 30 dead when the Zambian Air Force de Havilland Buffalo they were on crashed into the sea off Libreville, Gabon, after an engine fire.

Cyprien Ntaryamira and Juvenal Habyarimana

6 April 1994

The President of Burundi and the President of Rwanda, both Hutus, were among ten people onboard a Dassault Falcon 50. They were returning from a meeting of East and Central African leaders in Tanzania, the purpose of which was to discuss ways of ending ethnic violence in Burundi and Rwanda. As the plane approached the Rwandan capital of Kigali, it was shot down by rocket fire. No one ever established who was responsible for the attack, but the assassination of the first ever democratically elected Hutu leader sparked off the Rwandan genocide during which, only a couple of months later, the Rwandan military and extremist Hutus massacred at least 800 000 Tutsis and Hutu moderates. Eventually over a million people died in the slaughter.

John Denver

in bad weather in an overloaded plane. The Cessna 177B Cardinal stalled and crashed just after taking off from Cheyenne, Wyoming, killing Reid, Lloyd and Jessica.

Although Joe Reid was actually manipulating the controls of the aeroplane at the time of the crash, in the wake of the incident there was a change in US Federal law. It now prohibits record attempts by people who do not at least hold a private pilot certificate and a current medical certificate from being allowed to manipulate the controls of an aircraft 'during any record attempt, aeronautical competition, or aeronautical feat'.

Lieutenant Kara Spears Hultgreen

25 October 1994

As Hultgreen attempted to land her F-14 Tomcat onboard the aircraft carrier *SS Abraham Lincoln* off San Diego, California, her left engine stalled and she crashed into the Pacific. The death of the first US female pilot cleared for combat duty created a storm of controversy, but also seemed to inspire a profound change to a more positive attitude towards female fighter pilots.

Jessica Dubroff

11 April 1996

For reasons only known to her parents, Lloyd Dubroff and Lisa Blair Hathaway, seven-year-old trainee pilot Jessica Dubroff was attempting to be the youngest person ever to fly across America. Her flight instructor and official pilot-in-command Joe Reid, chose to take off

John Denver

12 October 1997

Country and western singer Henry John Deutschendorf Jr., better known as John Denver (53), crashed his experimental Rutang Long-EZ into Monterey Bay near Pacific Grove, California at 17:56. He lost control while reaching back to turn a fuel transfer switch. Denver's ongoing alcoholism and the determination that he was not even medically qualified to pilot the plane further marred the assessment of his death.

Michael Bell

2 April 1999

The New Zealand boxer was among all five who died when the South West Helicopters Aerospacial 350-B that he was flying in crashed into trees in the Rowallan Forest near Tuatapere. The cause was possible pilot 'incapacitation'.

John F Kennedy Junior

16 July 1999

The 38-year-old was piloting his Piper Sarartoga II off Martha's Vineyard, Massachusetts, when he lost control and the aircraft spiralled into the ocean. Kennedy, his pregnant wife Carolyn Bessette-Kennedy, their unborn child and his sister-in-law Lauren Bessette all died in the crash.

The incident is often held up as a seminal example of the difference of flying under VFR—visual flight rules—and IFR—instrument flight rules. Under VRF the pilots are responsible for avoiding objects that they can actually see. Obviously, this only works in conditions of good visibility. In haze, fog, storms or or similarly poor conditions, pilots rely on their instruments and operate under IFR rules. The catch is that they need to learn to understand and correctly interpret what their instruments are telling them. This takes training that leads to appropriate IFR qualifications. Even cautious, experienced IFR certified pilots are wary of flying in bad weather.

John F Kennedy Junior decided to fly in hazy conditions and at night, despite not being IFR qualified, having only had 310 hours of flying experience and despite his flying instructor offering to help him. The Transport Safety Board's subsequent expensive investigation determined that the Piper had no equipment malfunction and that the accident was purely the result of pilot error. The Kennedy family later made a large, undisclosed payment to the Bessette family.

Aaliyah Dana Haughton

25 August 2001

The 22-year-old singer's potentially phenomenal career and life ended when the Cessna 402B she was in crashed at the end of the Marsh Harbour, Bahamas,

runway after failing to take off. All of the nine onboard were killed. The pilot, Luis Morales III had obtained his licence under false pretences when he lied about his flying experience. An autopsy also revealed alcohol and cocaine in his blood. No one had informed the passengers that the plane was also grossly overloaded; the luggage hadn't even been weighed. In the aftermath of all this incompetence, Aaliyah's parents filed a wrongful death lawsuit against the carrier Blackhawk International Airways. The company later settled out of the court for an undisclosed amount.

Graham 'Shirley' Strachan

29 August 2001

The veteran Australian rocker and former lead singer of the 1970s band Skyhooks was on only his second solo flight as a pilot of a Bell 47 helicopter when he lost control in stormy weather. Around 16:00 he flew into strong air currents at the base of Mount Archer near Kilkoy, Queensland, which turned his helicopter upside down. In the crash he broke his neck and died instantly.

Cory Lidle

11 October 2006

The New York Yankees pitcher had only been flying for seven months when he and his Californian flight instructor, Tyler Stanger, had problems while flying under VFR over downtown New York. Lidle crashed a Cirrus SR20 into the Belaire Apartments in Manhattan's Upper East, wrecking the apartment of Kathleen Caronna, who wasn't home at the time. In the subsequent fire, three apartments valued at well over a $1 million each were damaged. The crash caused injuries to ten bystanders and to eleven of the 168 fire fighters who attended the scene. Both pilot and instructor died instantly.

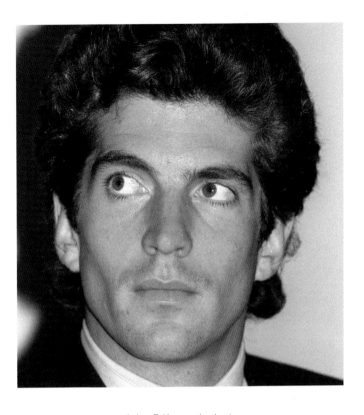

John F Kennedy Junior

Posters of Aaliyah Dana Haughton become larger-than-life shrines for her mourners.

IN THESE EARLY DECADES of humankind's forays into space, few people have yet to venture into the vacuum that surrounds our planet. Space travel is an extremely complex and hazardous undertaking, so much so that you can't even obtain private insurance coverage for astronauts. Many astronauts come from the ranks of test pilots, and are trained to the highest degree, but they're not supermen and superwomen. They are as vulnerable and mortal as anyone else.

PART SIX:
Deaths in space

In truth, no one has actually died in outer space itself, and the total number of human beings who have died in spacecraft has been low, but, when rockets and space shuttles are involved, the events make world headlines. Here now we compare the more spectacular accidents with the more down-to-earth, but no less tragic, deaths of astronauts closer to Earth.

23 March 1961

Casualty: Second Lieutenant Valentin Vasiliyevich Bondarenko (24)

In an accident vaguely reminiscent of the tragedy that would later claim the crew of *Apollo 1*, Bondarenko died during a simulation. In a capsule containing an atmosphere of 100 per cent oxygen, he was cleaning patches of biosensor equipment from his body with a cotton ball soaked in alcohol. He threw away the ball, without looking, in the direction of an electric stove hot plate. The ball ignited and set fire to his suit. A technician attempted to open the capsule door, but cabin pressure had effectively sealed it shut. He died sixteen hours later of shock, but for propaganda reasons the world didn't learn about his death until the 1980s.

28 February 1966

Casualties:
Captain Charles 'Charlie' Bassett (35)
Elliot McKay See (39)

Both astronauts died when their T-38 fighter crashed into the roof of a building at the McDonnell Aircraft Corporation plant during an instrument landing that went wrong.

27 January 1967

Casualties:
Lieutenant Commander Roger Chaffee (32)
Lieutenant Colonel Virgil 'Gus' Grissom (41)
Lieutenant Edward Higgens White (37)

Although technically not an air or space death, this accident shocked the world into realising that space travel could be dangerous even before the spacecraft took off.

The real tragedy of *Apollo 1*, or as the people concerned called it Command Module 012, was that it was caused by changes that NASA had made because of two earlier accidents.

In April 1960 NASA was testing the life-support system for the Mercury missions. They were using an oxygen nitrogen mix when too much nitrogen went into the spacesuit of test pilot GB North. North nearly asphyxiated, so NASA decided that from then on they would use pure oxygen.

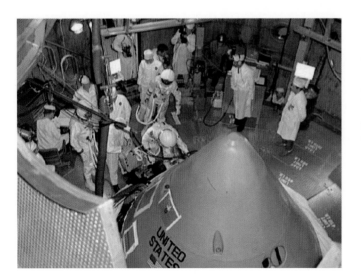

Astronauts entering the *Apollo 1* capsule.

The charred *Apollo 1* capsule after the fatal fire.

Columbia, the world's first spaceworthy shuttle being escorted by a Northrop T-38, the supersonic jet trainer in which so many astronauts—and potential astronauts—have lost their lives.

Then, on 21 July 1961 Virgil 'Gus' Grissolm became the second American in space during the suborbital flight of the Mercury Redstone-4, the *Liberty Bell.* After Gus splashed down the hatch blew off before the navy frogmen reach him, and the capsule flooded and sank. There was some debate as to whether Gus had accidentally triggered the hatch's explosive bolts, or whether, as Gus later claimed, the door blew off on its own because of a fault. NASA decided that from then on capsule doors would open inwardly, and not too easily, because a door that exploded out in space would empty the spacecraft's air and kill any astronaut who wasn't fully suited.

Now it was 1967 and North American Aviation (NAA) had built the *Apollo* capsule under the directive of NASA, but NAA had warned the space agency that a fully oxygenated capsule with a closed door could be inviting disaster. NASA chose to ignore these warnings. They also ignored the warnings of another contractor, General Electric. A GE executive, Hillard Paige, had even written NASA a letter, which stated bluntly that, 'The first fire in a spacecraft may well be fatal.' However, sixteen fire-free *Mercury* and *Gemini* missions in a pure oxygen atmosphere without a single capsule fire had made NASA officials complacent.

The crew of what was supposed to be the first *Apollo* Earth to Moon mission entered a command module for a test simulation. The three men, including Gus Grissom, were in Command Module 012, 65 metres (215 feet) high, on top of the three stages of the *Saturn V,* the largest rocket ever built. Unknown to anyone the insulation had become abraded around a silver-plated copper wire running through an environmental control unit.

'The interior air pressure became so high that it burst the capsule.
As technicians desperately tried to save the crew,
explosions and fire drove them back.'

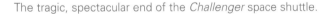

The tragic, spectacular end of the *Challenger* space shuttle.

Four hours into the simulation of the exhaustive pre-take-off checklist, the eroded wire sparked, igniting a leaking ethylene glycol (anti-freeze) cooling line. There was an immediate flash fire in the atmosphere of 100 per cent oxygen. At 06:31, Chaffee, in the pilot's chair, said simply: 'We've got a fire in the cockpit'. For a few seconds Chaffee attempted to transmit emergency messages, but his transmissions ended in a cry of pain. The air pressure in the capsule rose so high and so quickly that the inward-opening door was sealed shut. Unable to escape, the crew was burned alive. The interior air pressure became so high that it burst the capsule. As technicians desperately tried to save the crew, explosions and fire drove them back.

In the aftermath of the fiasco, heads rolled, including

Joe Shea, NASA's *Apollo* Program Manager. NASA 'reassigned' him to Washington to the role of Deputy Associate Administrator for manned space flight —a dead-end job if ever there was one. He suffered a nervous breakdown and left NASA after six months. He did return in the early 1990s as head of Space Station redesign and worked on the servicing mission for the *Hubble Space Telescope*.

The widows of the *Apollo 1* crew sued North American Aviation for the *Apollo 1* fire and eventually settled out of court.

The American Government posthumously awarded all three men the Congressional Space Medal of Honour. Several American schools are named after them.

24 April 1967

Casualty:

Colonel Vladimir Mikhailovich Komarov (40)

Cosmonaut Komarov died when his re-entry vehicle crashed to the ground after its parachute lines became tangled.

5 October 1967

Casualty:

Major Clifton Curtis 'CC' Williams (35)

A mechanical failure in the control systems of another T-38 fighter caused William's plane to crash near Tallahassee, Florida.

8 December 1967

Casualty:

Major Robert Lawrence (32)

Crashed his F-104 Starfighter while performing manoeuvres near Edwards Air Force Base, California.

27 March 1968

Casualty:

Colonel Yuri Gagarin (34)

Yuri Gagarin was the first man in space and the first man to orbit the Earth in *Vostok 3KA-2* on 12 April 1961. He and his flight instructor died during a routine training flight near Kirzhach, 48 kilometres (30 miles) east of Moscow. The cause remains uncertain.

A TRAGIC MILESTONE

Vladimir Komarov became the first person to die while actually on a space mission.

30 June 1971

Casualties:

Lieutenant Colonel Georgi Dobrovolski (43)
Viktor Patsayev (38)
Vladislav Volkov (35)

The first successful attempt to dock with the world's first space station *Salyut 1* was a welcome technical and publicity coup after the failure of *Soyuz 10*. From 7 June *Soyuz 11*'s crew spent 22 days in space, encountering several glitches along the way including an onboard fire. However, *Soyuz 11*'s record-setting triumph was short-lived.

Although the re-entry of the craft seemed normal to outside observers, when the recovery crew opened the capsule they found the crew dead from asphyxiation. In spite of heroic efforts, rescuers failed to revive them. Later investigations concluded that the cosmonauts had been dead for around 15 minutes. A valve located underneath the cosmonauts' seats, designed to equalise pressure between the cabin and the outside, had fired prematurely and the cabin's atmosphere had leaked into space through a hole only one millimetre in diameter. With the hole in such an inaccessible position, the cosmonauts were unable to locate and plug the leak before it was too late.

As a result of the accident, designers extensively reconfigured the *Soyuz* capsule to allow only two cosmonauts to ride, but it at least allowed enough room for them to wear space suits.

28 January 1986

Casualties:

Gregory Jarvis (41)

Sharon Christa McAuliffe (37)

Dr Ronald McNair (35)

Lieutenant Ellison Onizuka (39)

Judith 'JR' Resnik (36)

Major Francis 'Dick' Scobee (46)

Captain Michael Smith (40)

It was supposed to be a public relations triumph. The *Space Shuttle Challenger* was supposed to be an 'every man and woman' mission with a crew representing, in microcosm, the citizenry of America—female, male, black, white, Asian, military and civilian, Catholic, Protestant and Jewish. It was supposed to be an opportunity to show how safe space travel had become. So routine, in fact, that the major television networks

The *Challenger's* destruction wasn't quite complete. One of its booster rockets continues upwards, virtually undamaged.

weren't even going to bother to transmit live mission 51-L—the 25th shuttle mission.

Not everyone was so optimistic. For the past three years, engineers at Morton-Thiokol had been warning NASA administration about design problems with the solid rocket boosters (SRBs)—especially the O-ring seals, which held together joins in the SRBs. The seals were small, only 0.71 cm (0.27 inches) thick, and they were made of a synthetic rubber called Viton. Although small, they were sturdy and, because of their flexibility, designed to make up for tiny variations in the surfaces of the parts they joined together. Within their manufacturing tolerances they can withstand hundreds of thousands of times atmospheric pressure at sea level.

However, the freezing temperatures on that morning made Morton-Thiokol engineer Roger Boisjoly and his colleagues extremely worried. He knew that the rings had never been designed to operate at such low temperatures. On the night before the launch, engineers had been involved in a teleconference call between NASA and Morton-Thiokol's offices in Utah, Florida and Alabama. Morton-Thiokol's solid rocket manager and company representative Allan McDonald said that he would not sign the launch recommendation. The boosters needed to be at least 11.6 degrees Celsius (57 degrees Fahrenheit) or the O-rings wouldn't seal properly.

NASA booster manager Larry Mulloy dismissed the engineers' concerns as unfounded and unproven. He concluded the teleconference with the demand: 'My God, Thiokol, when do you want us to launch, next April?'

The browbeating worked. After 30 minutes Thiokol called NASA back and told them that as Allan McDonald wouldn't sign the recommendation to launch then booster program vice-president Joe Kilminster would.

Even up to the very last minute, the Morton-Thiokol engineers were warning NASA not to go ahead with the launch before inspecting the integrity of the rings and the seals. NASA had performed an inspection only the night before. The left booster was at 0.5 degrees

Celsius (33 degrees Fahrenheit) and the right booster was at −7.2 degrees Celsius (19 degrees Fahrenheit), yet NASA failed to regard the low temperature of the seals as significant. NASA ignored Morton-Thiokol as they had ignored them for the past three years. A Presidential Investigating Committee in June 1986 later found that NASA management was somewhat 'lackadaisical'.

'Even up to the very last minute, the Morton-Thiokol engineers were warning NASA not to go ahead with the launch before inspecting the integrity of the rings and the seals.'

Feeling under pressure—the shuttle had already been delayed for four days on account of weather problems and mechanical troubles—NASA management went ahead with the launch. There was still some excitement. Mission 25 would be the first time that an ordinary civilian, a teacher no less, would fly in the shuttle. Christa McAuliffe's class had come all the way from New Hampshire to watch the launch.

NASA cleared the shuttle to launch at 11:38.

At 0.678 seconds (T+0.678) after take-off, puffs of dark smoke emerged from the right solid rocket boosters (SRB). The puffs are at the O-ring seal joining the lower two sections of the SRB and near a strut holding the SRB to the central external tank.

In a normal take-off, the struts hold the SRBs until the SRBs run out of fuel and detach. The shuttle then continues to climb with the external tank (ET) in place until all the fuel is spent. Eighteen seconds after the shuttle's main engines shut down the ET detaches, falls to earth and breaks up in the atmosphere. By this time the shuttle achieves its orbit and the mission proceeds as planned.

At T+2.733 the last puff of smoke emerged from the O-ring join. At T+3.375 the last puff of smoke emerged from the area of the strut. Considering how long these puffs lasted no one on the ground noticed them at the

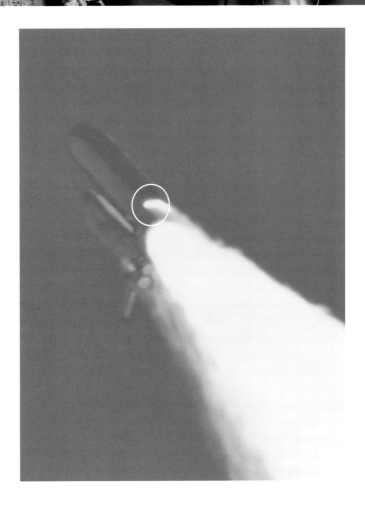

Shortly before *Challenger* exploded, flames erupted from the degraded O-ring seal (circled).

time. It was only on examination of video capture later that the true picture emerged.

At T+28, the shuttle power decreased as planned to control its ascent. However, the cold rings hadn't expanded quickly enough to seal the join in the right SRB. Hot gases escaped, burning the rings. The molten rubber and aluminium oxides form an unstable seal. Had that seal held for just two minutes, long enough for the SRBs to do their work and detach from the shuttle, the shuttle might have made it.

At T+40 and a height of 19 000 feet (5800 metres), the shuttle achieved a speed of Mach 1—equivalent to 1225 kilometres per hour (761 miles per hour). For the next thirty seconds high winds jostled the shuttle, putting the whole precarious arrangement under even more stress. The unstable, compromised seal in the right SRB began to fail.

T+58.788: ignited gas burned through a hole in the failed seal. The hole widened; the plume of fire grew to 12 metres and started to act like a blowtorch, burning through the sway strut that held the lower portion of the right SRB in place and burned a hole in the right aft part of the ET.

T+64.660: the shuttle's hull was breached. At this point computers were compensating for the abnormality and neither the crew or ground control detected anything wrong.

T+72.284: fourteen seconds of blowtorching finished its job and the lower strut failed. A quarter of a second later the right SRB pivoted on its upper strut and tilted its nose through the top of the external tank. The tank collapsed and tore it apart. The last thing anyone heard of the shuttle crew is Pilot Michael Smith saying, 'Uh, Oh'.

T+73.162: at 48 000 feet (14 kilometres) the *Challenger* disintegrated while the more robust SRBs detached and veered off in different directions. For the next 37 seconds they continued to fly until ground control ordered their deliberate self-destruction. The risk that they would fly into a populated area was too great. Controllers were left with the painful reality that they had to destroy vital evidence in order to preserve lives.

Roger Boisjoly watched, paralysed with shock and disbelief. He would spend the rest of the day in his office, unable to speak.

In a surreal moment the mission's public affairs officer Steve Nesbitt hadn't even noticed anything and was reading off data consoles. Millions around the world heard him continue to comment on the launch as if nothing had happened for several seconds after the explosion, while the cameras lingered on the burning

The remains of *Challenger's* external tank.

The surreal image of *Challenger's* demise is burned into the memory of millions around the world.

vapour ball in the sky. It gave some the impression that NASA was so blasé it wasn't even watching. Only a little later did he say: 'Obviously a major malfunction … the vehicle has exploded.'

The crew cabin continued to climb for a further five kilometres due to the forces of inertia. Its trajectory peaked at 65 000 feet (20 kilometres). Later examination determined that some of the crew survived the initial disaster—three or four had managed to activate their personal egress air packs, similar to the air masks that fall in a commercial airliner in the case of a depressurisation. It was possibly a futile gesture. The air from the packs was not under pressure, and if the cabin depressurised, there wouldn't have been enough air to keep the crew conscious during an almost three minute fall. In any event, the crew cabin hit the water at Cape Canaveral at 333 kilometres per hour (207 miles per hour)—a deceleration of 200 G. The cabin and its crew disintegrated on impact.

It would appear that the crew were doomed from the moment that the ignition sequence began. There was no escape scenario designed into the shuttle at that time. Of all the debris later recovered, the only human remains workers found was a small piece of bone and some tissue. NASA returned other identifiable items to the families. Dick Scobee and Michael Smith have graves in Arlington Cemetery and all the shuttle crew have a memorial at Arlington. The American Government awarded all of the crew with posthumous Congressional Space Medals of Honour.

The Presidential Commission on the Space Shuttle Challenger Accident—the Rogers Commission— included some of the most famous people involved in the space program, Neil Armstrong, Richard Feynman, Sally Ride and Chuck Yeager among them. Thousands of human-hours and millions of dollars later NASA had to undergo major structural and policy changes. The shuttle program was set back for years. The next launch would not be until 32 months later—with the *Discovery* launch on 29 September 1988.

What is left of the *Challenger* is now stored at Cape Canaveral Air Force Station in decommissioned missile silos. Fragments and debris still occasionally wash up on the Florida coast.

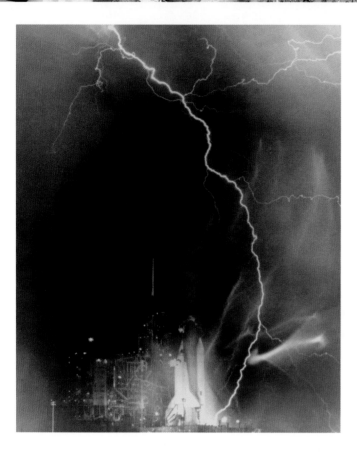

Lightning poses a risk to space ships as much as aeroplanes.

1 February 2003

Casualties:

Lieutenant Michael Anderson (42)

Captain David McDowell Brown (46)

Dr Kalpana 'Casey' Chawla (41)

Colonel Rick Husband (45)

Commander William 'Willie' McCool (40)

Commander Laurel Salton Clark (41)

Colonel Ilan Ramon (47)

As the *Columbia* space shuttle lifted off from Cape Canaveral on 16 January 2003, on the 113th shuttle mission, a piece of insulating foam broke off the external tank and damaged reinforced carbon-carbon panels and silicate thermal protection tiles near the leading edge of the shuttle's left wing. Foam loss had occurred on two of the previous missions, but hadn't caused any other problems.

When it was time for the shuttle to return to Earth, NASA said that they, 'took a very thorough look at the

24 May 1986

Casualty:

Stephen Thorne (33)

Thorne was a passenger on an Aerotech Pitts S-2A stunt plane when a short circuit electrical system distracted the pilot. The plane crashed near Santa Fe, Texas, killing both onboard.

17 June 1989

Casualty

David Griggs (49)

The astronaut crashed a North American AT-6D stunt plane in Earl, Arkansas. The determination was pilot error—overconfidence and improper planning.

Barring further accidents, the space shuttle program will have launched around 130 successful missions by 2010.

The official portrait of *Columbia*'s crew on the fateful 28th mission.

situation with the tile on the left wing and we have no concerns whatsoever. We haven't changed anything with respect to our trajectory design. It will be a nominal, standard trajectory'.

It wasn't. As the shuttle entered Earth's atmosphere, friction did its usual job of breaking the shuttle's descent, but the exposed part of the wing acted like an Achilles heel and the resulting, massive heat build-up spread to the whole wing.

Shuttle program manager Ron Dittemore later admitted that there was no contingency plan and they did not have the capability to repair the wing in orbit, even if NASA had concluded that it would be a problem.

Sixteen minutes before its due landing time, when the *Columbia* was above Texas at an altitude of almost 203 000 feet (62 kilometres), flight controllers in Houston lost communication with the *Columbia*.

At 14:04 GMT radar reports indicated a trail of debris behind the shuttle during re-entry. Heat and aerodynamic forces had compromised the left wing's structural integrity and the wing just came apart.

The crew lost control of the shuttle and the first fully operational space shuttle and veteran of 27 previous missions disintegrated, killing everyone onboard, including the first Israeli in a shuttle crew. The United States Government awarded all of the crew with posthumous Congressional Space Medals of Honour.

There would not be another shuttle flight for 30 months, until the launch of *Discovery* on 26 July 2005. It was to be the only shuttle flight that year. There are now only three shuttles in service: *Atlantis*, *Discovery* and *Endeavour*. Barring another accident there will have been from 131 to 133 shuttle missions until NASA discontinues the program in January 2010.

PART SEVEN:
Narrow escapes, near misses and close calls

AN INCIDENT DOESN'T ALWAYS have to turn into an accident; occurrences don't have to lead to fatalities. Sometimes by a combination of sheer skill and a lot of luck, aircrew can heroically pull victory out of the jaws of defeat. On other occasions, and for reasons that sometimes remain unclear, crews create catastrophic defeat out of otherwise normal events.

However, if an emergency happens it can be of some comfort to know that sometimes, despite the odds against it, some people do survive the seemingly unsurvivable.

8 October 1947

Flight: American Airlines Flight 311
Model: Douglas DC-4
Registration: NC90432
Fatalities: 0/54 (none of the 49 passengers or 5 crew)
Principle cause: practical joke

A gust lock is a device designed to disengage the controls, so that if there's a sudden movement due to turbulence pilots won't accidentally move the controls.

It was 07:55 and Flight 311 was flying at an altitude of 8000 feet (2440 metres) in clear weather over El Paso, Texas, en route from Dallas to Los Angeles. There were four people in the cockpit, the pilot, co-pilot, a navigator and another pilot hitching a lift in the jump seat. Neither the command pilot nor the jump seat pilot were strapped into their chairs. The jump seat pilot decided to pull a practical joke and secretly engaged the plane's gust lock. The command pilot tried to engage the elevator trim tab so that the plane could adjust to variations in pitch and keep a steady, level flight. With the gust lock on though, there was no response. When the jump seat pilot then switched the gust lock off, the trim tab engaged in such a way as to cause the plane to dive, roll and invert. The command and jump seat pilots fell out of their chairs and hit their heads on the control panel, accidentally feathering engines one, two and four. This caused the propeller blades to change angle and reduced the engines' thrust. The loss of power allowed the co-pilot to regain control of the plane and pull out of what would otherwise have been an irrecoverable dive.

'The jump seat pilot decided to pull a practical joke ...'

19 October 1953

Flight: Eastern Airlines Flight 627
Model: Lockheed 749A Constellation
Registration: N119A
Fatalities: 2/27 (2 of the 22 passengers and none of the 5 crew)
Principle cause: pilot error

At 00:56 Flight 627 was taking off from New York in foggy conditions, bound for San Juan, Puerto Rico. The plane was in the air for only a few seconds when the pilot became disoriented and crashed the plane onto the runway, whereupon it immediately burst into flames. What was exceptional about this crash was the heroism of the flight's chief purser and flight attendant who, at great risk to their own lives, went back into the flaming fuselage repeatedly in order to rescue all of the passengers they could. Because of their bravery only two passengers died.

28 June 1965

Flight: Pan Am Flight 843
Model: Boeing 707-321
Registration: N761PA
Fatalities: 0/153 (none of the 143 passengers or 10 crew)
Principle cause: engine disintegration

Flight 843 had just taken off from San Francisco en route to Honolulu and reached an altitude of 800 feet (244 metres) when the third stage turbine engine of engine four disintegrated explosively. The engine caught fire and the outboard fuel tank exploded, causing the separation of both the engine and the outer section of the right wing. The plane managed to make an emergency landing at Travis Air Force base.

Although the Constellation was a generally reliable aeroplane, three incidents in its first 10 months of commercial operation (1945–46) forced the authorities to temporarily ground the 'Connies' until Lockheed corrected design flaws that predisposed the engines to catching fire.

A cluster of McDonnell Douglas DC-8 jetliners docked around the San Francisco International Airport, California, about 1965.

Airforce personnel escorted the passengers and crew safely off the 707. Pan Am then dispatched another plane to pick them up, but then, in full view of the waiting passengers, as the relief plane landed its nose gear collapsed. It just wasn't their day.

22 November 1968

Flight: Japan Air Lines Flight 2
Model: McDonnell Douglas DC-8-62
Registration: JA8032, *Shiga*
Fatalities: 0/107 (none of the 96 passengers or 11 crew)
Principle cause: pilot error

The investigation later established that the flight crew was unfamiliar with the instrument flight director devices and the autopilot system. As the *Shiga* was making its final approach to San Francisco the crew miscalculated their approach and landed in San Francisco Bay 4 kilometres (2.5 miles) short of the runway.

Rescuers safely evacuated all of the 107 occupants. United Airlines recovered the *Shiga* 55 hours after the accident and repaired the damaged plane. The *Shiga* returned to Japan and remained in service for many decades after without further mishap.

It was the first time that a jet aeroplane had landed safely on water.

14 April 1970

Flight: Apollo 13

Apollo 11 had gripped the world's attention when Armstrong, Aldrin and Collins successfully achieved the first manned landing on the Moon. It remains one of the really big moments in human history. A little remembered historic footnote is that James Lovell and Fred Haise were on the backup crew for *Apollo 11*, and, had anything happened to the first team, Lovell would have been the first man on the moon.

Apollo 12 happened only a few months later when the Intrepid landed in the Ocean of Storms on 19 November. The work of Conrad, Bean and Gordon was only marginally less spectacular, but considerably more relaxed. Pete Conrad, who was rather shorter than Armstrong quipped before his step: 'Whoopee! Man, that may have been a small one for Neil, but that's a long one for me.'

To a certain extent, however, the Apollo missions were victims of their own success. By the time that *Apollo 13* launched on 11 April 1970, carrying James 'Shaky' Lovell, John 'Jack' Swigert and Fred 'Pecky' Haise, the public were growing a little blasé about the whole thing. The third mission in ten months didn't receive anywhere near the attention of the prior two.

That all changed in the early hours of 14 April when the spacecraft, comprising the command module

Odyssey, and the lunar module *Aquarius,* was 321 860 kilometres from Earth. Mission control asked the crew to stir the oxygen tanks to keep the liquid oxygen slush at an even consistency. When the crew applied power to the stirring motors, damaged Teflon-coated wires caught fire, heating the oxygen until the pressure caused the tank to explode. The explosion also damaged the remaining oxygen tank and the *Odyssey* was crippled.

Continuing the tradition of problematic sentences and misquotes during space missions, John Swigert said to ground control, 'OK, Houston. We've had a problem here.' James Lovell then reiterated, 'Houston, we've had a problem.' However, the press misquoted the phrases with the much pithier, and much more popular, 'Houston, we have a problem.'

The crew had to abort their mission and resort to a survival scenario that NASA had worked out only a little while before the launch. They used the *Aquarius* as a lifeboat, conserving what oxygen and power they had left for the vital adjustments they'd need to negotiate a slingshot orbit around the moon and re-enter the Earth's atmosphere.

NASA had worked out the lifeboat scenario for only two men for two days. Adapting the situation to allow three men for four days the astronauts had to live with considerable discomfort. One critical problem was the build up of carbon dioxide. NASA scientists devised a way of adapting CO_2 scrubbers from the *Odyssey* to make them work in the *Aquarius.* If they hadn't, the crew would have died from CO_2 poisoning.

In the absence of enough drinking water and problems dealing with human waste, the astronauts had to store their urine and faeces in the ship. Ejecting the matter might have caused changes to the spacecraft's trajectory. Fred Haise developed a urinary tract infection from the stress of it all.

In addition NASA ordered them to turn off the heating in order conserve power. In the cold, condensation from the astronauts' bodies began to build up in the cabin. Fortunately, on account of the extensive redesign of the electrical system, which NASA had instigated after the

Apollo 1 disaster, when the crew switched the systems on for re-entry the spacecraft was functioning perfectly.

The crew jettisoned the *Aquarius* in orbit and the craft that had served as the life raft for the astronauts burned up on re-entry. One part did survive, a power generator containing 3.9 kilograms (8.5 pounds) of plutonium. It's currently lying at the bottom of the sea, in the Tonga Trench, and will remain radioactive for the next 2000 years.

By the time that the *Odyssey* splashed down in Pacific, half way between the Cook Islands and Tonga, the crew of *Apollo 13* were cold, sick, exhausted and very, very smelly, but they were alive. It was 17 April 1970 and they had been in their spaceship for five days and almost 23 hours when the recovery crew from the USS *Iwo Jima* picked them up.

In the extensive analysis that followed, an irony revealed itself. Just over eight hours before the explosion, the number two oxygen tank gauges had shown levels off the scale. It was because of this misreading that NASA had asked the crew to stir the tanks more often than the mission had originally required. If it hadn't been for this original fault Swigert might have had the explosion while Lovell and Haise were on the Moon's surface. If that had happened, none of the men would have been able to return.

In a final irony the mission did have one major success. Because of the lunar flyby *Apollo 13* flew further away from Earth than any other mission. It is a record that it has held now for almost 40 years.

The *Apollo 13* missions operations team and the crew were awarded the Presidential Medal of Freedom for their efforts.

In 1994 Jim Lovell and *TIME Magazine* journalist Jeffrey Kluger wrote *Lost Moon: The Perilous Voyage of Apollo 13*. It became the basis of the 1995 Ron Howard film. Technicians later rebuilt the *Odyssey* from parts that had survived re-entry and you can now see it on display at the Kansas Cosmosphere and Space Centre in Hutchinson, Kansas. In the final analysis NASA considers the *Apollo 13* mission to have been a 'successful failure'.

The crew of *Apollo 13* being picked up after splash down—miracles do happen.

Inside the cockpit of a BAC 111. Commercial pilots are highly trained, competent professionals who are generally confident and optimistic, but, then again, you'd have to be.

6 September 1971

Flight: Pan International Non-Scheduled Flight
Model: BAC-111 515 FB
Registration: D-ALAR
Fatalities: 22/121 (21 of 115 passengers and 1 of 6 crew)
Principle cause: maintenance error

Shortly after taking off from Hamburg, both of the plane's Rolls-Royce engines failed, forcing the plane to make an emergency landing on the Hamburg-Kiel Autobahn, near Hasloh. The plane collided with a bridge and both wings sheared off. The cause of the crash: a maintenance worker had filled the plane's water-cooling system with kerosene.

24 November 1971

Flight: Northwest Airlines Flight 305
Model: Boeing 727
Fatalities: 0/42
Principle cause: potential bomb

One of the most bizarre stories in the history of commercial aviation has since become the inspiration for quite a few theories and speculations. The facts, as far as we know them are these:

On the day before Thanksgiving, NA 305 had just taken off from Portland, Oregon, when a man seated in 18C handed an attendant a note, stating that he had a bomb and if he didn't receive $200 000 and four parachutes when the plane landed in Seattle he would blow up the plane.

The authorities chose to comply. At Seattle the man received his $200 000 (worth around $1 million today) and allowed the 36 passengers to leave as well as co-pilot and engineer, keeping only the pilot and the stewardess. He then demanded that the pilot take him to Mexico. The man, who called himself Dan Cooper, instructed the pilot to fly no higher than 10 000 feet (3000 metres) and no faster than 200 miles per hour (320 kilometres per hour). Somewhere over Ariel, Washington, in an area of hemlock forest north of Portland, Dan Cooper asked the flight attendant how to open the tail stairway exit. After asking her to move to the front of the plane he bailed out, carrying the 10-kilogram (21 pound) bag of money with him—a white-wheat coloured bank bag.

The temperature was below zero. Winds of 150 kilometres per hour (92 miles per hour) and freezing rain pelted the plane. Dan Cooper wasn't dressed

'Besides the books,
         rock songs and the film that Dan Cooper inspired,
                   no one ever heard from him again ...'

for the weather. Investigators assumed that he died either in the fall or of exposure.

In the aftermath of the incident, Boeing installed a device that prevents the tail exit stairs of a 727 from being opened in flight. Interestingly the device is called a 'Cooper Vane'.

The day after the jump the police investigated a Portland resident called DB Cooper, but soon cleared him. Unfortunately, on account of lazy journalists the newspapers continued to call the skyjacker 'DB Cooper' and the name stuck.

In spite of the massive and expensive five-month-long manhunt that followed, interviews with nutcases and spurious deathbed confessions, in all 1000 'serious suspects' later they never found a trace of the Dan Cooper the FBI described: 'mid-40s, 6 feet tall, 170 pounds, black hair, a bourbon drinker, a chain smoker'.

In 1980 an eight-year-old boy, Brian Ingram, was playing along the banks of the Columbia River digging in a fire pit when he chanced upon $5800 worth of $20 bills buried in the sand. The serial numbers on the bills matched those handed over to Cooper.

As well as the books, rock songs and the film that Dan Cooper inspired, such as the 1982 film, *The Pursuit of DB Cooper* starring Treat Williams and Elwood Reid's 2004 novel, *D.B.*, no one ever heard from him again ... unless you count the events of 1995.

In March of that year 70-year-old Florida antiques dealer Duane Weber lay dying in Pensacola Hospital when he whispered to his wife: 'I'm Dan Cooper.'

Jo Weber had been married to the secretive man for seventeen years and didn't have a clue what he was talking about, but Duane said that he'd sustained his old right-knee injury 'jumping out of the plane' and further claimed that he'd forgotten where he'd buried $173 000 in a bucket. He died eleven days later.

After Duane's death, Jo found that Duane was 47 at the time of the skyjacking and fitted the FBI's description of Dan Cooper very closely. Even Duane's handwriting matched a sample in a book on Cooper. Initially the FBI treated Jo like another crank, but she eventually made contact with Ralph Himmelsbach, who had headed the investigation between 1971 and his retirement in 1980, and he pulled strings to reopen the investigation. The evidence was circumstantial and inconclusive, but still compelling.

In 1990 Jo had found a wheat-coloured bag hidden in Duane's van. In 1994 she had found an old plane ticket for a Northwest Airline trip from Seattle to Tacoma, but couldn't find the ticket again after Duane's death. Duane already knew of his disease in 1971, thought that he wouldn't live past 50 and was in the process of divorcing his fifth wife. At the time he was a desperate man.

The FBI closed its investigation of Duane Weber in 1998 citing the need for more conclusive evidence. The case remains the only officially unsolved skyjacking. It could have been worse. At least no one was hurt, unless you count Duane's knee.

244

PART SEVEN | Narrow escapes, near misses and close calls

A Northwest Airlines pilot communicates on a cockpit telephone. Note the condition of the paint and rivets. You can't judge a plane by its cover, however. Northwest has suffered only 18 fatalities in over 80 years of operation.

SOLE SURVIVORS

Sometimes miracles happen. In the midst of horrendous accidents and huge loss of life, the odd lucky soul comes through.

Juliane Koepcke

Eight years after Pan Am 214 was struck by lightning, killing everyone onboard and under similar weather conditions, Lineas Aereas Nacionales Sociedad Anonima (LANSA) Flight 508 was flying from Lima to Pucallpa, Peru. The plane was carrying 86 passengers and six crew when it flew into severe thunderstorms at an altitude of 21 000 feet. At 12:36 on Christmas Eve, 1971, lightning struck the right wing, which separated, then part of the left wing, dooming the aircraft, its six crew and 85 of its passengers.

Miraculously, seventeen-year-old Juliane Koepcke remained strapped to her seat as the plane disintegrated all around her. She fell 3 kilometres (2 miles) into the canopy of the Amazon Rainforest and sustained only a break to her collarbone, a gash in her right arm and the loss of eyesight in one eye.

Even more miraculously as many as fourteen other people may also have survived the fall, but only Juliane was able to find help in time. Uniquely prepared with jungle survival skills she had learned from her father, Juliane found a river and waded through knee-high water on and off for nine days, following the water downstream, in the hope of finding civilisation.

Eventually she came across a canoe shack, where local loggers found her and tended her wounds. She was also covered in insect infestations, which the men cleaned using salt and kerosene. The following day, after a seven-hour canoe ride, they arrived at a logging station in Tournavista, where she was airlifted to hospital in Pucallpa and reunited with her father.

Juliane was the only survivor of Flight 508. Her mother Maria Koepcke died in the plane. Whoever else might have survived the initial explosion and fall didn't make it out of the jungle. LANSA Flight 508 holds the world record for the greatest number of fatalities in an air accident due to lightning.

German film director Werner Herzog had missed Flight 508 because of over-booking and made a documentary about Juliane in 2000, called *Wings of Hope*. Juliane is now back in Germany and works as a zoologist.

Vesna Vulovic

On 26 January 1972, one month after Juliane Koepcke's ordeal, terrorists from the Utashe group—fighting for an 'ethnically pure' Croatia—planted a bomb in the forward cargo hold of Jugoslovenski Aerotransport (JAT) Flight 364. The McDonnell Douglas DC-9-32 was on its way to Beograd Airport, Serbia from Copenhagen, Denmark when, over Srbska Kamenice in Czechoslovakia, the bomb exploded and tore the plane apart. Six crew and 27 passengers died almost immediately.

Flight attendant Vesna Vulovic hadn't originally been meant to be on that flight and was only there because of a mix-up with another attendant called Vesna.

Nevertheless, strapped to her seat, she fell 33 000 feet (10 kilometres) for three minutes before hitting a snow-covered mountain. A German man, who had been a medic during World War II, found the wreckage of the plane and all of the passengers dead, except for Vesna. Her injuries included a fractured skull, two broken legs and three broken vertebrae. Doctors told her parents that she would not survive and she was in a coma for three days before regaining consciousness. Her injuries left her paralysed from the waist down for several months. Doctors attributed her survival to her low blood pressure, which prevented her heart from bursting at high altitude. Paul McCartney later presented her with an award from the *Guinness Book of World Records* for surviving the highest fall without a parachute—a record that is unlikely ever to be beaten.

Sometimes survival statistics defy logic. On 5 September 2005, when Mandala Airlines Flight 091 crashed into a heavily populated suburb of Medan in North Sumatra, 109 died, but 16 passengers survived. Another 39 people on the ground were killed.

The only survivor of the 23 passengers and four crew of JAT Flight 364, Vesna became a national hero in Yugoslavia in the 1970s and later continued working for JAT—at a desk job. In an interview with Philip Baum in 2005—over thirty years after the crash—she had this to say: 'I'm not lucky. Everybody thinks I am lucky, but they are mistaken. If I were lucky I would never had this accident and my mother and father would be alive.

The accident ruined their lives too.'

The explosion and fall were so terrible that Vesna's mind refused to keep any of the memories. She remembers, 'Nothing … The last thing I remember is boarding the plane by the rear door and seeing a few women cleaning the plane … The next thing I can remember is seeing my parents in the hospital.'

In spite of having the largest route network and being the world's sixth largest airline, Delta has an enviable safety record.

'... strapped to her seat, she fell 33 000 feet (10 kilometres)
for three minutes before hitting a snow-covered mountain.'

OTHER SOLE SURVIVORS

21 January 1985: On a Galaxy Airlines charter flight, the aeroplane started vibrating after take-off from Reno because maintenance operators had failed to shut a door properly. Thinking the problem was with the engines, the pilot cut power but lost so much airspeed that the plane fell out of the sky. Of the 65 passengers and six crew, three people initially survived, but two died within a month. The sole survivor was a seventeen-year-old boy, thrown from the wreckage when the plane hit the ground.

16 August 1987: From the very beginning there were bad omens with Northwest Airlines Flight 255. During the taxi in Detroit, the pilot missed the correct turnoff. Investigation later concluded that they also failed to go through the proper take-off procedure. If they had, they might have noticed that the wing flaps and slats weren't properly configured. They might also have picked up that there was no power to the plane's take-off warning system. During take-off the plane rolled, failed to gain enough altitude, struck two lampposts and a building, slid into a railway embankment upside down and burst into flames, incinerating all its occupants except four-year-old Cecilia Chichan, who lost her mother in the crash.

20 November 1993: An Avioimpex Flight flew into Mount Trojani, Macedonia, killing 115 out of the 116 passengers and all eight of the crew.

11 January 1995: Of the 47 passengers and five crew on Intercontinental de Aviacion Flight 256, only one nine-year-old girl survived when the DC-9 crashed in a marsh after electrical problems affected the altimeter.

3 September 1997: All six crew members and 59 out of the 60 passengers on a Vietnam Airlines Tupolev-134B died when it crashed in heavy rain at Phnom Penh, Cambodia.

15 December 1997: Only one of the eight crew members and none of the 77 passengers survived the crash of a Tajikistan Airlines Tupolev 154B-1 when the plane crashed 13 kilometres (8 miles) short of Sharjah in the United Arab Emirates.

6 March 2003: Only one of the six crew members and none of the 97 passengers survived the crash of an Air Algérie Boeing 737-200 soon after one of the plane's engines caught fire during take-off from Tamanrasset, Algeria.

8 July 2003: Three-year-old Mohamed al Fateh was the only person to escape from the wreckage of a Sudan Airways 737-200C when it crashed east of Port Sudan, Sudan. Eleven crew and 105 passengers died, including the boy's mother. Mohamed suffered major injuries and also lost his right leg.

27 August 2006: Only one of the three crew and none of the 47 passengers survived the crash of a Comair CRJ-200ER shortly after take-off from Lexington, Kentucky.

3 November 1973

Flight: National Airlines Flight 27
Model: McDonnell Douglas DC-10-10
Registration: N60NA
Fatalities: 1/128 (1 of the 116 passengers and none of the 12 crew)
Principle cause: pilot 'experiment'

While flying near Socorro, New Mexico, en route from Houston to Las Vegas, the captain and flight engineer tried an experiment. The instruments that measured the rotational speed of each engine's low-pressure compressors provided information to control the plane's auto throttle. What would happen if they tripped the circuit breakers? They soon found out. In the absence of any feedback information, the starboard engine suddenly began to over-speed. The engine began to go into a 'multi-wave high speed resonance'—in short it started to vibrate to the point where the fan blade tips made contact with the engine cases. At thousands of revolutions per minute, the fan blade assembly disintegrated. Shards of the blades travelling at high velocity speared the right wing, the fuselage and entered engines one and two. One fragment penetrated and broke a window, causing an explosive cabin decompression. The passenger who had been sitting next to the window was violently sucked out the hole.

Nineteen minutes later the plane made a safe emergency landing at Albuquerque, New Mexico.

7 August 1975

Flight: Continental Airlines Flight 426
Model: Boeing 727-224
Registration: N88777
Fatalities: 0/138 (none of the 131 passengers or 7 crew)
Principle cause: windshear

Within seconds of take-off from Denver, Colorado, the 727 experienced severe windshear. Having only attained a height of 100 feet (30 metres) the plane crashed at the end of the runway. No one died, but the plane was so badly damaged it had to be written off.

9 February 1982

Flight: Japan Airlines Flight 350
Model: McDonnell Douglas DC-8-61
Registration: JA8061
Fatalities: 24/174 (24 out of the 166 passengers and none of the 8 crew)
Principle cause: mental instability

Flight 350 was just about to land at Tokyo-Haneda Airport. The landing gear and flaps were down and the plane was travelling at 280 kilometres per hour (174 miles per hour) when the co-pilot called out '500 feet'.

For the next 25 seconds the co-pilot worried that the pilot had not followed standard procedure and confirmed the landing. They were approaching the decision height of 200 feet (60 metres), the point at which the flight crew either went through with the landing or, for whatever reason, aborted. As there was nothing evidently wrong, there was no reason not to proceed smoothly with the landing. At 08:43:56 the altimeter warning sounded, as the flight engineer called out '200'. At that point the captain suddenly switched two of the engines into reverse, in an attempt to destroy the plane. A struggle ensued in which the co-pilot tried to regain control, but it was too late, and the DC-8 crashed into Tokyo Bay.

The pilot had recently come back from being off-duty for a year on account of mental illness. To date this remains one of only a handful of uncontested instances where a pilot's mental instability, unconnected to an act of terrorism, caused a fatal plane crash on a major commercial flight.

'... the captain suddenly switched two of the engines into reverse,
in an attempt to destroy the plane.
A struggle ensued in which the co-pilot tried to regain control,
but it was too late ...'

Indonesian police guard the wreckage of a Lion Air plane near Solo international airport, Central Java. At least 32 people were killed and 61 injured when the aircraft skidded off the runway after landing in bad weather on 30 November 2004.

25 May 1982

Flight: VASP Scheduled Passenger Flight
Model: Boeing 737-2A1
Registration: PP-SMY
Fatalities: 2/118 (2 out of the 112 passengers and none of the 6 crew)
Principle cause: pilot error

The pilot had used a rain repellent on the plane's windshield, which changed the way that light refracted through the glass. It caused an optical illusion that made the ground appear to be further away and at a different angle to what it actually was. As the plane was landing in rainy conditions, the pilot miscalculated the plane's height and angle and the 737 made a hard landing, nose gear first. The gear collapsed, the jet skidded nose down on the runway and finally broke in two.

24 June 1982

Flight: British Airways Flight 009
Model: Boeing 747-236B
Registration: G-BDXH
Fatalities: 0/262 (none of the 247 passengers and none of the 15 crew)
Principle cause: volcanic eruption

British Airways 009 was flying between Kuala Lumpur and Perth and had reached a height of 37 000 feet (11 300 metres) when it flew into the plume of the eruption of Mount Galunggung. Smoke and dust penetrated the flight deck and there were even manifestations of ball lightning (St Elmo's Fire). Volcanic ash and pitting made it almost impossible to see through the windshield and caused all four engines

One of the more impressive images from the history of near misses. If you've ever wondered if there was a crane big enough to lift a 747—here's your answer.

to fail. The plane began to lose altitude rapidly, but the crew managed to restart engine four at an altitude of 13 000 feet (4000 metres). The other engines soon came online but continual surging in engine two forced the crew to shut it down again. Fortunately, the jumbo was able to make a safe emergency three-engine landing at Jakarta.

10 March 1984

Flight: Union de Transports Aériens (UTA) Scheduled Flight
Model: McDonnell Douglas DC-8-63PF
Registration: F-BoII
Fatalities: 0/23 (none of the 18 passengers or 5 crew)
Principle cause: bomb

In a prelude to a much more serious sabotage five years later (see page 72 [UTA 772]) the DC-8 was standing on the tarmac for a one-hour stopover in Brazzaville, in the Congo, when a small bomb exploded in the baggage compartment. The passengers were evacuated and twenty minutes later another bomb in the baggage compartment exploded and destroyed the aircraft.

Even though his subjective weight fluctuated from near weightlessness to five times normal during the fall, the captain managed to regain control

19 February 1985

Flight: China Airlines Flight 006
Model: Boeing 747-SP-09
Registration: N4522V
Fatalities: 0/274 (none of 251 passengers or 23 crew)
Principle cause: engine failure

At 10:15 while en route to Taipei at an altitude of 41 000 feet (12 500 metres), 550 kilometres (344 miles) northwest of San Francisco, California, the number four engine of Flight 006 suddenly lost power. The plane rolled right, nose down and began an uncontrollable descent.

While the plane was falling, the occupants of the 747 experienced vertical acceleration forces from −4 to +5.1 G. Even though his subjective weight fluctuated from near weightlessness to five times normal during the fall, the captain managed to regain control when the plane reached 9500 feet (2900 metres).

The plane suffered major structural damage, but only two passengers were seriously injured.

Flight 006 made it back to San Francisco but, adding insult to injury, while landing the crew forced the landing gear through closed doors. The doors broke off and damaged the tail and elevators. The subsequent inquiry criticised the captain's performance, citing his:

… preoccupation with an in-flight malfunction and his failure to monitor properly the aeroplane's flight instruments, which resulted in his losing control of the aeroplane. Contributing to the accident was the captain's over-reliance on the autopilot after the loss of thrust on the No. 4 engine.

3 February 1988

Flight: American Airlines Flight 132
Model: McDonnell Douglas DC-9
Registration: N569AA
Fatalities: 0/131 (none of 120 passengers or 11 crew)
Principle cause: undeclared hazardous chemicals

It was very lucky that the incident occurred as Flight 132 was on its final approach to Nashville, Tennessee. Unknown to its crew—because the items had been undeclared and improperly packed—the DC-9 was carrying a drum of hazardous materials in its mid-cargo compartment, near the wing area.

The item was a 47-kilogram (104-pound) drum containing two chemicals used in textile treatment—sodium orthosilicate and hydrogen peroxide solution. The peroxide leaked into the orthosilicate and caused the chemicals to catch fire. Within three minutes, at 16:03, a light smoke was rising from the floor near seat 17E. The plane touched down at Nashville twelve minutes later, and authorities evacuated passengers on the taxiway. Fire fighters successfully managed to extinguish the fire, but not before the aircraft suffered substantial damage.

28 April 1988

Flight: Aloha Airlines Flight 243
Model: Boeing 737-297
Registration: N73711
Fatalities: 1/95 (no passengers and 1 of 5 crew)
Principle cause: airline and FAA negligence

This was an accident that could have been so much worse. The investigation determined that there were several factors that contributed to it: Aloha's maintenance program failed to notice disbonding and metal fatigue in a lap joint in the fuselage of the 737; Aloha's management failed to supervise its maintenance crew properly; and the FAA had failed in its duty of care to assess Aloha's maintenance procedures, especially considering that Aloha planes were constantly being subjected to pressurisation stresses in a humid, salt air climate.

The end result of this negligence was that one afternoon while on a short run from Hilo, Hawaii, to Honolulu and flying at 24 000 feet (7300 metres) near Maui, a lap joint in the 737 failed. A 6-metre (18-foot) section of the fuselage ripped out of the main body. In the explosive decompression one stewardess was sucked out. Her body was never recovered.

Fortunately she was the only casualty. The crew of Flight 243 successfully negotiated an emergency forced landing at Maui airport, albeit 75 kilometres per hour (46 miles per hour) faster than a normal landing.

26 June 1988

Flight: Air France Demonstration
Model: Airbus A320
Registration: F-GFKC
Fatalities: 3/136 (3 of 130 passengers, none of 6 crew)
Principle cause: pilot error

FROM THE COCKPIT

Captain addressing the passengers: 'Ladies and Gentleman ... We shall take off for a short tourist flight starting at the Habsheim Flying Club where we will do two fly-overs to demonstrate the quality of French aviation.'

It was supposed to be a public relations triumph, but it turned into a fiasco. Scores of journalists were in the new Airbus flying over the airfield at Mulhouse-Habsheim. The beginning of the air show demonstration flight started without incident at 14:41. The low fly-over just four minutes later was only supposed to go down to 100 feet (30 metres), but the pilot took it down to only 35 feet (9 metres) and seemed to apply recovery power a little too late.

The plane was too low to avoid a forest at the end of the runway. The A320 crashed so badly that it was written off. Miraculously, only three people died in the impact. The survivors, mostly press, didn't have very flattering things to say about the 'quality of French aviation'.

Amazingly, considering its explosive decompression, this Aloha Airlines 737 made an emergency landing—in convertible mode —on the Hawaiian island of Maui. One flight attendant died and 65 of the 90 passengers were injured. The psychological cost is a little harder to calculate.

10 June 1990

Flight: British Airways Flight 5390

Model: BAC-111 528FL

Registration: G-BJRT

Fatalities: 0/93 (none of the 87 passengers or 6 crew)

Principle cause: maintenance error

Just over a day before the flight, maintenance crews incorrectly replaced the left windscreen of the plane. The investigation later determined that out of the 90 securing bolts, 84 were smaller than the specified diameter and the remaining 6 were too short.

British Airways 5390 was at 17 300 feet (5275 metres) en route from Birmingham, England, to Malaga, Spain.

The cockpit of the Dutch Martinair DC-10 after the plane crashed and burst into flames at Portugal's Faro airport on 21 December 1992. Of the 340 passengers 50 died and more than 200 were hospitalised.

At 07:33, only thirteen minutes into the flight, the bolts failed under the effects of internal cabin air pressure and the windscreen blew out initiating a massive decompression in the cockpit. The flight deck door blew in and moisture in the air suddenly condensed. Anything loose in the passenger compartment flew towards the front of the plane.

Pilot Tim Lancaster was immediately sucked out of his chair and was careering head first to certain death, when his knees caught on to the flight controls just long enough for flight attendant Nigel Ogden to grab him. If Nigel hadn't been in the flight deck at the time, Lancaster would have been sucked out—the forces were equivalent to those on an ant under a vacuum cleaner.

> 'Pilot Tim Lancaster ... was careering head first to certain death, when his knees caught on to the flight controls just long enough for flight attendant Nigel Ogden to grab him.'

Co-pilot Alistair Atchison took control of the plane and commenced an emergency descent. The lives of everyone onboard were in his hands. In the rushing air Atchison couldn't hear ATC's responses to his maydays, as a result, there was some delay in putting emergency services in place.

The plane was now nose down and other flight crew rushed to relieve Ogden, who was beginning to suffer from exhaustion and frostbite as he struggled to maintain a grip on Lancaster, who was slipping

further and further down the nose of the plane. Meanwhile, Atchison secured an emergency landing at Southampton.

For the remaining 22 minutes of the flight and through all of the jostling and bumping of the emergency landing the cabin crew were able to hold on to their captain. As the passengers were evacuated, emergency crews rescued Tim Lancaster, who suffered from frostbite, fractures to his right arm, right wrist and left thumb and extensive bruising.

Apart from Nigel Ogden's more minor injuries, no one else was hurt.

29 April 1993

Flight: Continental Airlines / Jetlink Flight 2733
Model: Embraer EMB-120 RT Brasilia
Registration: N24706
Fatalities: 0/30 (none of the 27 passengers or 3 crew)
Principle cause: pilot error

Although some can take their jobs a little too seriously others may have a tendency to be a little too relaxed.

The Brasilia had taken off from Little Rock, Arkansas, at 15:16 bound for Houston. In overcast sky with light winds the weather wasn't ideal for flying, but it wasn't that bad. The incident that was to follow was completely avoidable.

While the Brasilia was climbing, the co-pilot wrote in his log book and ate his meal while the captain put his feet up on the console, loosened his safety harness and relaxed. When the plane reached 15 800 feet (4800 metres), the flight attendant was in a hurry to start serving the meals, so she asked the captain to take the plane to cruising altitude faster so that the plane could level out. Being easygoing he complied by setting the controls and putting the plane on autopilot.

Over the next few minutes the plane's pitch doubled from a safe and sensible 3.2 degrees to a rather more daring 6.4 degrees and the rate of climb more than doubled from 420 feet (130 metres) per minute to 900 feet (275 metres) per minute. Unfortunately, this reduced the airspeed from 320 kilometres per hour (200 miles per hour) to 261 kilometres per hour (162 kilometres per hour). Meanwhile ice was starting to accumulate on the wings.

At 15:33, at 17 000 feet (5200 metres) the plane stalled, immediately disconnecting the autopilot. The Brasilia began to roll left and right as much as 90 degrees while pitching down as much as 67 degrees. The airspeed climbed to 390 kilometres per hour (240 miles per hour) as the plane dived at a rate of 17 000 feet (5200 metres) per minute. The captain assumed that the left engine was racing. In reality the shaking had damaged it, but shutting it down and lowering the landing gear helped to put the Brasilia back under control at 5500 feet (1680 metres). The captain declared an emergency and headed for Pine Bluff. In the approach the Brasilia overshot and touched earth with only 575

258

FROM THE COCKPIT

11:45:15—Captain: I hate droning around visual at night, in weather without having some clue where I am.
11:45:23—Co-Pilot: Yeah, but the longer we go out here the …
11:45:24—Captain: Yeah, I know.
…
11:49:13—Captain: This is a can of worms.
…
11:50:23—Flight 1420 Cockpit: We're down … we're sliding.

metres (1880 feet) of runway remaining. By now it was raining and the wet wheels skidded and hydroplaned off the tarmac and onto wet, sodding at the end of the runway. The NTSB report didn't mince words:

The captain's failure to maintain professional cockpit discipline, his consequent inattention to flight instruments and ice accretion, and his selection of an improper auto flight vertical mode, all of which led to an aerodynamic stall, loss of control and a forced landing. Factors contributing to the accident were poor crew discipline, including flight crew coordination before the stall and the flight crew's inappropriate actions to recover from the loss of control. Also contributing to the accident was fatigue induced by the flight crew's failure to properly manage provided rest periods. Fortunately no one was hurt.

1 June 1999

Flight: American Airlines Flight 1420
Model: McDonnell Douglas MD-82
Registration: N215AA
Fatalities: 11/145 (10 of 139 passengers and 1 of 6 crew)
Principle cause: pilot error

Flight 1420 had left Dallas very late. By 11:20 it was 30 minutes from its ultimate destination at Little Rock Arkansas—two hours behind schedule. The weather was deteriorating and two huge storm systems were flanking the plane and steadily closing in. The captain chose to fly into the narrow corridor between the storms instead of diverting to another airport even though the winds were gusting at Little Rock as fast as 140 kilometres per hour (87 miles per hour)—far beyond the recommended maximum for the MD-82.

Investigators examine the wreckage of American Airlines Flight 1420 that crashed while trying to land in stormy weather at Little Rock National Airport. Eleven people were confirmed dead and 86 injured in the crash.

At 11:39 ATC suggested a change of approach which added another ten minutes to the flight time. Experts later concurred that the flight crew of 1420 were suffering from a bout of severe 'get-there-itis'. In fact the NTSB enquiry later revealed that pilots were routinely flying into thunderstorms and that there were no clear guidelines for alternative decisions. The pilot of Flight 1420 even decided to use visual flight rules to save time, even though the huge storms had reduced visibility to dangerous levels.

As they made their landing the flight crew made one more critical error. They failed to arm their spoilers at the right time to reduce their altitude and speed. When they engaged reverse thrust, they didn't lose enough speed, and the plane landed hard and much too fast. They hit the runway at over 160 kilometres per hour (100 miles per hour) and the plane started shaking violently. Sixty-four tons (140 000 pounds) of aircraft was now aquaplaning completely out of control.

The MD-82 ran off the end of the runway, crashed through a security fence and down a rock embankment before finally slamming into a steel walkway, only metres from the Arkansas River.

The NTSB spent 18 months investigating the crash before concluding that the combination of bad weather and worse judgement calls had cost the captain his life and that of, miraculously, only ten of his passengers. The MD-82 was a complete write off.

23 September 1999

Flight: Qantas Flight 1
Model: Boeing 747-438
Registration: VH-OJH
Fatalities: 0/429 (none of 410 passengers or 19 crew)
Principle cause: crew miscommunication

QF 1 is Qantas Airlines' seminal route—London to Sydney. It, therefore, seems poetically appropriate that the most serious incident in the jet history of the world's safest airline should happen while QF 1 was landing at Bangkok on the first 'hop' of its sentimental 'kangaroo route'.

The problem occurred seconds before landing, during the heaviest rain that the first and second officer had ever experienced during an approach.

The 747 approached the runway at the upper limits of both its speed and height. At 22:46:27 the rain was so heavy that the crew could only see the landing guide lights briefly between each swipe of the windscreen wipers. It was dark and the captain was concerned about the plane's speed and the fact that 750 metres along the runway the plane was still 10 feet (3 metres) off the ground. He ordered the co-pilot to 'go around', professional shorthand for 'abort landing, take the plane up and try again'.

In retrospect it still seems a sensible decision; although the Qantas crew didn't know it, a Thai Airways Airbus 330 captain had made the same call three minutes earlier. The co-pilot started the manoeuvre manually rather than engaging the more instantaneous automatic take-off/go around system. His intention was to engage it a few seconds later.

Suddenly, three seconds after the Qantas captain had made his call, the weather cleared and the plane touched down at a point about 1002 metres down the runway. The captain aborted the go around; however, the plane was now going too fast for a safe landing and there was also confusion about who was in charge and what to do.

The crew engaged the landing gear brakes, but it didn't occur to them to apply full reverse thrust. Although the plane slowed down, it didn't slow fast enough. The wheels aquaplaned on the ungrooved tarmac. The plane overshot the runway and ran onto a boggy grass field, jarring so violently that cabin ceiling panels were dislodged, and came to rest with its nose on an airport perimeter road. The whole incident had taken 41 seconds.

Although there were 38 minor injuries, the

A rare position for a Qantas plane to find itself in. The 747 slid off Bangkok Airport's runway while landing in heavy rain.

biggest complaint from the passengers was that the precautionary evacuation did not commence until 20 minutes after the landing. The absence of a fire may have contributed to the lack of immediate urgency.

Qantas has never publicly confirmed how much the accident cost them, but some have quoted figures as high as $100 million.

22 December 2001

Flight: American Airlines Flight 63
Model: Boeing 767-323
Fatalities: 0/197 (none of 185 passengers or 12 crew)
Principle cause: hijacking

It must have seemed like a sick joke to the occupants of American Airlines Flight 63. Just over three months after the 9/11 attack on the World Trade Centre, here was another hijacking.

Richard Reid had attempted to board this same Miami-bound flight the day before, but authorities at Charles de Gaulle Airport had stopped the UK citizen. Why they let him on the next day is anyone's guess.

Shortly after a meal service, while they were flying near Boston, passengers smelled smoke in the cabin.

Flight attendant Hermis Moutardier found Reid attempting to light a match. She told him that smoking wasn't allowed on flights. He promised to stop, but was trying again several minutes later. This time Hermis asked him, 'What are you doing?' and he lunged at her, revealing that a shoe on his lap had

PART SEVEN | Narrow escapes, near misses and close calls

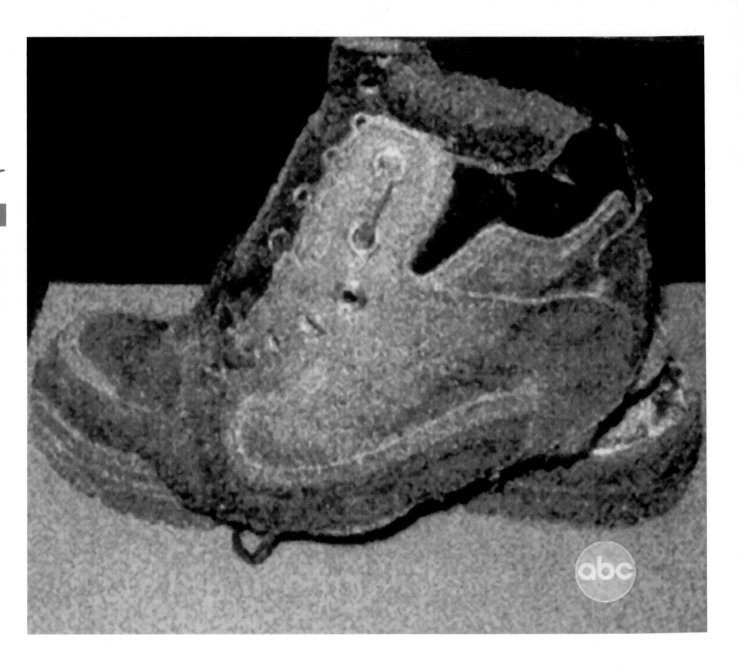

One of the shoes worn by 'shoe bomber' Richard Reid. The amount of plastic explosives hidden in his shoes could have blown a hole in the fuselage of the American Airlines flight from Paris to Miami. He was overpowered by flight attendants and passengers while trying to ignite the shoes.

a fuse leading into it. As she tried to stop him, he pushed her onto the floor and Hermis started screaming. Flight attendant Christina Jones attempted to grab Reid and he bit her thumb. Other cabin crew then joined the struggle.

Eventually the passengers and crew managed to control the man using a combination of plastic handcuffs, seatbelt extensions and headphone cord.

Better prepared since the World Trade Centre attacks, two fighter jets escorted Flight 63 to its emergency landing at Logan International Airport. Authorities parked the 767 in the middle of the runway and bussed the passengers to the main terminal after arresting Reid.

Examination of Reid's shoes revealed a combination of over 100 grams of pentaerythrite tetranitrate and triacetone triperoxide explosives in the hollowed out soles of his black basketball shoes—potentially enough to blow a hole in the plane.

An American Federal Court in Boston found Reid guilty of terrorism on 30 January 2003. In combination with other charges his sentences add up consecutively to three life terms plus 50 years and, rather redundantly, five years of supervised release. He's also liable for fines and restitutions totalling $2 006 882.17—to be exact.

It wasn't until 27 March 2006 though that Zacharias Moussaoui—one of the candidates for 9/11's '20th Bomber'—revealed that he and Reid had intended to hijack a plane and crash it into the White House on 9/11 2001. Although there is some doubt about the truth of this claim there's no doubt that Reid is currently serving his sentences, in Florence, Colorado, where he lives in mostly solitary confinement with concrete furniture and under 24/7/365 closed-circuit video surveillance.

WORDS OF A WITNESS

In an interview on CNN the next day, Miles O'Brien spoke to passenger Kwame James about the incident:

James: At first I thought someone was having a seizure or something, but then a flight attendant approached me and said, 'We need some help.' So I ran back there and we started wrestling with him ... me and about three or four other guys ... and he was just, I mean, unbelievably strong. We held him down then eventually a doctor came and gave him an injection ... to try to subdue him and stuff. And we pretty much had to tie him up with belts and everything we could get our hands on to tie someone up.

O'Brien: How many people were involved in subduing him, Mr. James? And, as I understand it, you're a professional basketball player. I assume you have a little bit of height. For you to say it was difficult to subdue him, I suspect he must have been a very powerful person.

James: Yes, I would say he was about 6-4, 220. And I'm about 6-8, 220 to 225, and it took me and three or four other guys just to hold him down. He almost seemed possessed.

O'Brien: Did you hear him say anything before, during or after this whole incident?

James: When we got him kind of calmed down, we ... me and the other guys ... asked him a couple questions like, 'Why this?' and all that stuff. And he told us, 'We'll all see.'

SURVIVORS' STORIES

There are several factors that make it difficult for those who haven't lived through an air disaster to know what it is actually like:

- Air crashes are statistically rare. You're much more likely to meet someone who's been involved in a car crash.
- When air crashes happen it's a fine line between a serious incident and a major disaster. There's little inbetween. You're much more likely to meet someone who's lived through a car crash than someone who's lived through a plane crash.
- When you're in the middle of a serious plane incident the experience is so terrifying that you're in an altered state of consciousness.
- Those who do survive aren't necessarily great storytellers.
- Those who do survive don't necessarily want to tell the story because it can bring up horrific and painful memories. It's difficult to justify going through all that just to satisfy the curious. Not everyone feels that it's 'helpful to talk about it'.

There's nothing like being there and frankly, most of us wouldn't want to be. We should be grateful for whatever insights we can find from those who have survived and who are willing to share what they lived through and others didn't.

29 September 2006

Captain Joseph Lepore (42) and First Officer Jan Paul Paladino (34) had logged over 14 400 flying hours between them when the brand new $25 million Legacy 600 they were flying collided in midair with Gol Flight 1907 at 16:48. The Gol 737-800 crashed eleven minutes later with the loss of all onboard, but the Legacy sustained only relatively minor damage.

Nevertheless, it was enough to put the small business jet in an emergency situation. The passengers included two employees of the manufacturer, Embraer, and two employees of the jet's new owners, ExcelAire Service. The fifth passenger was Joe Sharkey. Fortunately for the history of aviation accidents, Sharkey was the ideal witness—a business travel columnist doing a special report on corporate jets for *Business Jet Traveller Magazine*.

In interviews and in an article in *The New York Times* dated 3 October 2006, 'Colliding With Death at 37,000 Feet, and Living', Sharkey describes the flight as uneventful and comfortable when, 'Without warning, I felt a terrific jolt and heard a loud bang, followed by an eerie silence, save for the hum of the engines.'

Looking out the window Sharkey noticed that the left winglet was gone and that, 'The leading edge of the wing was losing rivets, and starting to peel back.'

Although no one panicked, the plane kept losing speed. As the minutes dragged on the passengers began to write notes to the people they cared about, hoping that if the worst happened the notes would survive the crash.

It took Captain Lepore 25 minutes of searching and calling before he announced, 'I can see an airport.' Co-Pilot Paladino later said, 'We didn't know how much runway we had or what was on it.'

Sharkey then related that they 'came down hard and fast. I watched the pilots wrestle the aircraft because so many of their automatic controls were blown. They brought us to a halt with plenty of runway left. We staggered to the exit.

Even after they had made it safely to ground, the pilots and the passengers didn't know what had hit them, or what they had hit. It wasn't until 19:30 that Embraer executive Dan Bachman, who spoke Portuguese, heard the story. Captain Lepore said that, 'If anybody should have gone down it should have been us.'

Brazilian aviation authorities and police questioned the crew and passengers of the Legacy at the airbase over a period of 36 hours following the crash. The day after the accident a military plane took the seven men to the police headquarters in Cuiaba. Part of the reason for the intensity of the investigation might have been that, up until that point, according to Sharkey, the Brigadeiro Velloso airbase was supposed to be secret.

Lawyers for the families of the crash victims have filed suits against ExcelAire and Honeywell, the manufacturers of the transponders. However, there remains the revelation that on two separate occasions ATCs in Brazil instructed the Legacy to maintain its altitude at 37 000 feet. This has put considerable doubt on the liability of the smaller plane's crew and operator.

6 March 1978

On 16 November 2006 I conducted an interview with Christopher Squibb. His story comes from a different time and place, but it is incredibly revealing and brutally honest. Christopher's narrative demonstrates that extreme situations tend to bring out extreme

The wreckage of the Air France Airbus A340 at Toronto's Pearson International Airport on 3 August 2005. It slid off the runway and burst into flames. All 297 passengers and 12 crew onboard miraculously escaped largely unscathed from the blazing aircraft.

behaviour, and whatever veneer of civilisation that we like to believe we have tends to evaporate when we're confronted by raw, naked fear and the overwhelming need simply to survive.

Christopher: You won't have heard about what happened on our flight because it was in 1977 and at the time Gulf Air wasn't a part of IATA [International Air Transport Association] so what happened was hushed up. We weren't allowed to talk about it to anyone.

Xavier: How did they manage to cover it up?

Christopher: The Arabs worked to their own rules.

I was working as a flight attendant. I was 21 and I'd only been flying for a year. It was 6 March 1978. We were

The smouldering wreckage of Garuda Airlines Boeing 737-400—carrying 140 people including Australian journalists and diplomats—which overshot the runway and burst into flames on 7 March 2007.

on GF 2, London to Bahrain via Paris, on a Lockheed-1011 TriStar. Gulf Air hadn't been flying that long, but the plane wasn't new because GA tended to buy their aircraft from other carriers like British Airways ...

It was about 22:00 and we were taxiing down the runway about to leave Charles de Gaulle Airport. From where I was sitting I was the only crew member who could see the whole port side of the plane ...

... we were taxiing down the runway, going full speed, with the maximum thrust that you need to get off the ground. We were a nanosecond from take-off when a crew member in front of me who was travelling civilian notices this wall of flame out the window and says, 'There's a fire.' Then I could see it too. I could see flames coming up from the right wing. I reached for the phone to talk to the cockpit when I realised that that was useless. When you're taking off, there are so many checks and procedures that the flight crew have to go through that answering the phone is the last thing on their minds. You just don't do it. So my only choice was to get up and go to the flight deck ...

I quickly glanced outside and I noticed that we were just leaving the ground. Just as I reached the flight cabin door and I was opening my mouth to say that there was a fire on the wing there was this almighty explosion and the plane just dropped. Fortunately, it was almost the perfect situation. Everything was secured. The trolleys were secured. Everyone except me was strapped in. It wasn't like those situations where people were standing up and walking around ...

The plane just careened down the runway. It just fishtailed, swaying back and forth at a tremendous speed. I don't remember what runway we were on but it was a long one, which was lucky.

Then all hell broke loose. The oxygen masks all fell out and people started screaming.

Xavier: Was there any time to think?

Christopher: I think that when something truly horrible happens, then everything happens in slow motion. Being inside a plane that is out of control is just the most frightening experience. It's just so big. People get scared just in turbulence, but when a plane is shaking like that, it feels as if the interior cabin is shaking independently of the outer fuselage. The engines are roaring the engines are at full reverse. The alarms are going. It's horrible. I kept thinking, 'We're going to blow up, or we're going to crash. We're not going to make it.' I was petrified, but I was also on automatic pilot. The guy who had trained us was ex-SAS. He told me one thing I've never forgotten. He said, 'Don't think you know how you'll react in an emergency. I've seen big, strong men wet their pants and I've seen petite women come to the fore.' And he was right. The French stewardesses we had were just amazingly calm, competent and in control, and the Arab boys we had on crew ... just lost it totally.

To come to a standstill seemed like an eternity. It was probably the longest, what, 30 seconds of my life.

The evacuation sirens started going off ... Pandemonium just happened. You realise then that there's no such thing as ever evacuating a plane unless it's at a standstill.

Human beings react.

Xavier: There's no opportunity to get people under control?

Christopher: You think that you would, but you don't, because people have such a survival instinct ... It was first in best dressed. The seats just concertinaed forward and people were just climbing over the top of them. And we did get to our doors to tell people where to go, but people were just hysterical ... You think that you'll do the right thing but you don't ...

The doors all opened and the chutes started inflating, except for the back two doors didn't have chutes installed, they had life rafts instead, but in the panic some ten people were just pushed out of the back doors and fell about 25 feet [8 metres] to the ground ... One of the people who was pushed out was a woman who was seven months pregnant. She lost the baby.

To make matters worse the centre left door, didn't have a chute either. I was on the opposite door getting people down the slide. In some cases I had to push them because they didn't want to go. What's supposed to happen is that the first couple of people down the slide are 'able bodied', and they're supposed to hold down the inflatable chute slide as other people come down, but people just ran and the chute was just flapping in the wind ... People didn't want to get on this flapping thing. Some people were taking their baggage. It was crazy, they just didn't want to listen, but we'd evacuated nearly all of the people off the plane before the fire trucks arrived ...

We had a load of about 250 passengers and about fifteen crew and we got everyone off in about one minute and 35 seconds ... But while I was standing at that doorway, gripping the handle and getting people off,

everything was still in slow motion and it seemed like an eternity. I truly thought that we were all going to die. It wasn't until I was out of the plane myself that I felt that I might just live.

The moment people were on the ground, the airport authorities just started taking people away. Then there'd be this sudden wave of hostility. I guess it was some sort of knee-jerk reaction. Some of them actually wanted to get back in. They figured that they'd survived, so they were now worried about their passports and valuables. I couldn't believe it. I mean, you nearly die and you start worrying about your papers? There were armed police having to move people away. I mean, *the plane was still on fire.*

The crew all ended up in this conference room at the airport and we were just locked away for two days with access only to food and toilets. We were told that they didn't want the passengers and the press talking to us.

The official version of the events was that the wheel had caught fire, rubber had spun up into the engine and the engine had caught fire, but that was bullshit. Wheels just don't catch fire. Unofficially, the rumour was that it was that someone had planted a bomb in the wheelbase. That's what we thought at the time anyway.

While we were in this room, the only way we were able to make a call out was because one of the girls was French and she found a phone in a storeroom and we managed to get a line out so that we could call our families and tell them that we were okay ...

After about *36 hours* we were taken to London and put into a hotel room for a couple of days. We were then taken back to Bahrain and led into an office one by one. I was told that I could have a ticket to anywhere and I was going today. I went to Australia.

> 'One of the people who was pushed out
> was a woman who was seven months pregnant.
> She lost the baby.'

Firefighters battle flames during an aircraft fire training exercise.

The whole thing left a really nasty taste. A lot of crew left after that because they felt that they were just numbers, that if anything ever happened we'd never be told anything. There were no work laws, no unions, nothing.

Xavier: What about you?

Christopher: You know, prior to the event I'd been in turbulence and I banged my head on the ceiling when the plane dropped 500 feet [186 metres]. I'd been in a plane that had to land with only one of its three engines running. I had a really cavalier attitude. Now, I hate flying. It terrifies me. Every little shake, every little noise makes me crazy. I'm a basket case if the crew goes into the flight deck for any reason. I was at Heathrow, boarding a plane an hour after 9/11 happened. A girl in front of me just fainted. I was dragged onto the plane kicking and screaming. I was a mess. I had to be tranquilised. What helped was that it was a Garuda flight and I figured that terrorists wouldn't attack fellow Muslims. When I get on a plane now all I pray is that I'll have an uneventful flight.

WHEN MANY PEOPLE HEAR THE WORDS 'plane crash', the first thing that pops into their heads is the image of an aircraft falling to ground, plummeting into the sea or, ever since 9/11, smashing into a building. As valid as these images are, they don't always include one of the most horrifying scenarios—fire.

Cynical observers have noted that planes are not designed for survivability, and true pessimists are quick to point out that a considerable proportion of a commercial passenger jet's mass is the wings full of fuel. Planes also contain hydraulic liquids under tremendous pressure and literally thousands of metres of electrical wiring. Onboard fires have been responsible for some of the worst disasters in aviation history and it's a miracle of design and maintenance that planes don't catch fire more often.

PART EIGHT: 08
Fire: the passenger's worst nightmare

And that's just half the story.

Even if a plane survives a forced landing, there are many cases where the real killer wasn't the crash itself, but the ensuing fire. The idea of being roasted alive is horrific, and it's not much comfort to know that you're actually more likely to die from smoke inhalation or the lack of oxygen. Burning jet fuel, fabrics and other materials tend to create a greasy, acrid, foul-smelling smoke that blinds you, chokes you and makes you vomit.

It seems odd then that, even now, the airline industry is not making any great steps to make jet fuel safer for passengers. Then again, the whole history of air disasters is full of examples of profits coming before people.

14 August 1972

Flight: Interflug Non-Scheduled Passenger Flight
Model: Ilyushin 62
Registration: DM-SEA
Fatalities: 156/156 (148 passengers and 8 crew)
Principle cause: fire

Many people had noticed that a hot-air tube in the rear of the plane had a leak. It had been like that for some time, but no one had done anything about it. The Ilyushin 62 was on its way from Berlin to Bourgas Airport in Bulgaria when air at a temperature of 300 degrees Celsius (572 degrees Fahrenheit) started blowing out from the compromised tube. The air weakened electrical insulation and aeroplane controls. Within seconds of take-off at 16:30 a short circuit spewed out extremely hot sparks into the rear of one of the cargo bays.

By 16:43 the plane was climbing through 29 000 feet (8900 metres) and the crew requested an emergency return to Berlin. At 16:51 the crew dumped fuel and at 16:54 they initiated an emergency descent. By now they were losing altitude and were aware that they had a fire in the rear of the plane. At 16:59 they reported the fire to ATC and issued a mayday. Two minutes later the fire had weakened the rear fuselage so much that the tail section fell off. The Ilyushin broke up in midair killing all onboard.

26 November 1979

Flight: Pakistan International Airlines Flight 740
Model: Boeing 707-340-C
Registration: AP-AWZ
Fatalities: 156/156 (145 passengers and 11 crew)
Principle cause: cockpit fire

Investigators were never able to determine the cause of the cockpit fire that brought down Flight 740. There was no evidence of sabotage or electrical failure and

in the end authorities considered that the most likely candidates were a leaking gasoline or kerosene stove, which Haj pilgrim passengers had carried onboard and had ignited with the differences in cabin air pressure.

We do know that at 01:29 PK740 left Jeddah, Saudi Arabia, on its way to Karachi, Pakistan, and was climbing when a flight attendant reported a fire near the rear passenger cabin door. The fire was intense and spread quickly; the passengers panicked. At 02:03 the pilot declared a mayday. That was the last transmission the crew of Flight 740 sent. At 02:04 the plane crashed on a plateau 48 kilometres (30 miles) north of Taif in Saudi Arabia.

19 August 1980

Flight: Saudi Arabian Airlines Flight 163
Model: Lockheed L-1011 TriStar
Registration: HZ-AHK
Fatalities: 301/301 (287 passengers and 14 crew)
Principle cause: fire

Flight 163 took off at 16:06 from Riyadh, Saudi Arabia, and was six minutes and fifty-four seconds into its flight to Jeddah when indicators warned that there was a fire in a rear cargo compartment. At 16:20 the crew began a return to Riyadh. By 16:22 there was a noticeable amount of smoke in the rear of the passenger cabin and the passengers began to panic. Three minutes later the fire had penetrated into the cabin and people were fighting in the aisles.

It was another eleven minutes before Flight 163 made its approach to Riyadh and it finally touched down at 18:36.

The investigators never determined the cause of the fire, but cited several contributory causes for the tragedy that followed:

- The failure of the captain to prepare the cabin crew for immediate evacuation upon landing and his failure in not making a maximum stop landing on the runway, with immediate evacuation;

- The failure of the captain to properly utilise his flight crew throughout the emergency;
- The failure of C/F/R (Crash, Fire, Rescue) headquarters management to ensure that its personnel had adequate equipment and training to function as required during an emergency.

Instead of shutting down the engines immediately and evacuating the plane, the pilot continued to taxi and didn't stop the engines until 18:42. Crash, fire and rescue teams waited for an evacuation. Unknown to them, an electrical failure compromised the door locks and the crew was desperately trying to open the doors from the inside. It was a twenty-minute struggle. When the emergency services finally made it through a door, they found nothing but a plane full of corpses, engulfed in flames and smoke.

22 August 1985

Flight: British Airtours Flight 328M
Model: Boeing 737-226
Registration: G-BGJL
Fatalities: 55/137 (53 out of the 131 passengers and 2 of the 6 crew)
Principle cause: fire

Flight 328M was a charter tour trip from Manchester, UK, to the Greek Island of Corfu. Shortly after 06:12 as the 737 was attempting take-off, the left engine combustion chamber cracked and flew apart. A piece of engine then pierced an underwing access panel. The captain heard a thud and immediately aborted the take-off.

At 06:12:45 the captain advised ATC that he was abandoning take-off, and as he was doing this, the fire alarm activated. ATC advised that they could clearly see the fire and that rescuers were on their way.

Emergency services remove a body from the wreckage of the British Airtours Flight 328M Boeing 737, which burst into flames during take-off at Manchester Airport. Fifty-five people lost their lives on the holiday flight bound for Corfu, but 82 survived.

At 06:13:01 the crew decided to make a starboard evacuation. Seconds later the captain advised the co-pilot to go into an access road to keep Manchester's one active runway clear for other planes. This touch of courtesy resulted in a twenty-second delay, which confirms the adage that 'no good deed goes unpunished'. As the plane stopped, the crew began an evacuation drill, but almost immediately a breeze blew flame from the engine onto the fuselage and the jet fuel ignited.

Smoke seeped through the left side windows as the fire penetrated the tail section, filling the cabin with thick, greasy black smoke. Both the captain and co-pilot escaped through the right cockpit window as the cabin crew tried to open the right front door—but the door had jammed.

Twenty-five seconds after the plane had stopped the purser opened the front left door. There were now 131 passengers trying to escape, and, although the attempted evacuation was reasonably calm, many were jammed in the cabin's narrow centre aisle. People stumbled. People were being trampled and as they gasped their first taste of burning plastic and jet fuel they involuntarily inhaled even more.

By now the purser had managed to open the front right door and passengers had managed to open the right overwing door and the rear door, but burning chemicals and thick smoke were blinding them. Few, if any had bothered to count the number of chairs between their seat and the nearest exit. If they had, they might have been able to feel their way out with more confidence and certainty. Many might not have walked right past the open overwing exit that they couldn't see. None in fact even used the right wing door.

Cyanide gas inhalation, carbon dioxide inhalation and smoke inhalation killed more than a third of the passengers.

Firefighters spray water onto the remains of an Indian Alliance Air Boeing 737, which crashed 17 July 2000, into a housing estate in the eastern Indian town of Patna. All 6 crew members and 55 out of the 58 passengers died in the crash along with a further 5 ground casualties—the result of a stall due to crew error.

9 May 1987

Flight: LOT Polskie Linie Lotnicze Flight 5055
Model: Ilyushin 62MK
Registration: SP-LBG
Fatalities: 183/183 (172 passengers and 11 crew)
Principle cause: fire

The 62MKs have four rear-mounted engines below the tail. Flight 5055 left Warsaw at 10:18 bound for New York. At 10:41 a turbine shaft in number two engine broke loose and the turbine disc failed. Debris punctured the fuselage and started a fire in the cargo hold and severed the elevator controls, forcing the crew into an emergency descent and into shutting down number one engine as well. The crew then mistakenly thought that in shutting down the engines that they had also extinguished the fire. Although an emergency landing at Modlin Airport would have been quicker, they decided to turn back and land at Warsaw because of better rescue equipment there. At 11:09 the fire had run wild for 28 minutes, and, as Flight 5055 made a left turn for its final approach, the crew lost control. At 11:09 the plane crashed into a forest 6 kilometres (4 miles) from Warsaw at a speed of 465 kilometres per hour (289 miles per hour).

28 November 1987

Flight: South African Airways Flight 295
Model: Boeing 747-224B
Registration: ZS-SAS
Fatalities: 159/159 (140 passengers and 19 crew)
Principle cause: fire

Flight 295 was approaching Mauritius and nearing the end of its long flight from Taipei when a fire broke out in the cargo hold. Smoke from the fire not only disoriented and incapacitated the crew, but the heat and flames eventually burned through several control

cables and weakened the aircraft's structure to the point of deformation and ultimate midair collapse. The plane fragments hit the Indian Ocean 250 kilometres (156 miles) south-east of Mauritius at 00:07.

11 July 1991

Flight: Nationair Flight 2120
Model: McDonnell Douglas DC-8-61
Registration: C-GMXQ
Fatalities: 261/261 (247 passengers and 14 crew)
Principle cause: fire

It seems that Middle Eastern Airline flights carrying pilgrims on their Haj to and from Mecca suffer disproportionately from destruction by immolation.

Flight 2120 was flying pilgrims back home to Nigeria and had just taken off from Jeddah at 08:29. The crew reported that they were having trouble pressurising the cabin, when the plane reached around 2500 feet (760 metre). Two under-inflated tyres had failed during the first 500 feet (150 metres) of take-off and one of them had caught fire. Although the crew turned back almost immediately, the fire spread through several hydraulic lines making it impossible for the flight crew to control the elevators. As the plane approached Jeddah again to make an emergency landing the nose pitched down sharply and the plane crashed.

The accident was the worst ever involving a DC-8.

21 August 1995

Flight: Atlantic Southeast Flight 529
Model: Embraer EMB-120ER Brasilia
Registration: N256AS
Fatalities: 8/29 (7 of 26 passengers and 1 of 3 crew)
Principle cause: fire

Flight 529 is an example of how fire can hit a plane with a double whammy, and, while not a major accident in terms of lives lost, it is an outstanding example of skilful and courageous handling.

The Brasilia was a veteran of 18 171 flights and ASE's management had already scheduled it for its last flight. On the day that turned out to be its actual last flight, the plane was to fly between Atlanta, Georgia, and Biloxi, Mississippi. It took off at 12:25. Nineteen minutes into its flight the Brasilia was at 18 100 feet (5500 metres) and on its way to its cruising altitude of 24 000 feet (7300 metres). Suddenly there was a *bang* and the plane lost thrust from its left engine. A propeller had snapped off.

The enquiry later revealed that:

The fracture was caused by a fatigue crack from multiple corrosion pits that were not discovered by Hamilton Standard [the blade manufacturers] because of inadequate and ineffective corporate inspection and repair techniques, training, documentation and communication. Contributing to the accident was Hamilton Standard's and FAA's failure to require recurrent on-wing ultrasonic inspections for the affected propellers.

Propeller blades are hollow and units made from cork hold their core balancing weights in place. The corks were soaked in chlorine, which contributed to corrosion and fine fractures, which ultrasound inspection failed to pick up.

With the effective loss of its port engine the Brasilia immediately lurched left and started to dive. Aerodynamic forces peeled the outer skin off the engine. A fire began and the cabin filled with smoke as burning oil seeped into the air-conditioning intake.

After falling for 500 feet (150 metres), Captain Ed Gannaway and Co-Pilot Matt Warmerdam managed to regain some control, but by the time the plane's altitude

was 15 400 feet (4700 metres) the port engine had come loose from its wing mountings. This unbalanced the plane and interrupted airflow over the wing.

Fortunately, the plane was designed to fly on one engine, but the complications caused the plane to continue losing altitude as the crew attempted to return to Atlanta.

By the time the plane had reached 4400 feet (1350 metres), it was flying beneath clouds and was still 13 kilometres (8 miles) and 2 minutes away from Atlanta. Although the pilots' efforts were heroic they were running out of time. They couldn't see the airport and they had so many things going on that an instrument landing was impossible and they switched to visual flight rules.

At almost 12:53 Flight 529 crashed in a field just over the Alabama–Georgia border. Area resident Mrs Bill Jeters called 911 and said simply 'Yes, we have a plane crashed in our backyard.' In all the confusion no one at ATC had called emergency services.

Amazingly at this moment all 29 of the plane's occupants were still alive. Ed Gannaway was unconscious and in his seat. Matt Warmerdam's right shoulder was dislocated and in the confines of the crashed cockpit he was trying to beat a hole through the layered tempered glass and composite his windscreen with an axe. Aft, there was a huge hole in the plane and the more conscious and able passengers immediately scrambled to escape, but only a few made it out before the plane caught on fire. A passenger later related that, as the flames engulfed them, 'You could see some people whose flesh was dropping off their bodies or their faces. It was getting worse and worse.'

By 13:05 help arrived and everyone except the pilots were off the plane. Passenger David McCorkell along with former fire chief Steve Chadwick and his crew of Carroll County firefighters were doing their best to help the pilots before they burned to death.

Matt Warmerdam was not expecting to make it out alive and told Chadwick to: 'Tell my wife Amy that I love her.'

Chadwick replied, 'No Sir. You tell her you love her because I'm getting you out of here,' which he did.

The survivors were taken to Tanner Hospital, 15 miles away. The burn victims' injuries ranged from minor to 92 per cent of their bodies.

Later, many would honour the memory of Captain Gannaway—a hero who didn't make it.

While he had been trapped in the cockpit, an oxygen cylinder had burst behind Matt Warmerdam. It fed a raging fire that left burns to 42 per cent of his body, but he lived. Many months and skin grafts later he returned to flying, saying: 'It's what I do.'

Hamilton Standard completely overhauled its inspections and repairs and there have been no more fan blade failures since Flight 529.

11 May 1996

Flight: ValuJet Flight 592
Model: McDonnell Douglas DC-9-32
Registration: N-904VJ, *Critter*
Fatalities: 110/110 (105 passengers and 5 crew)
Principle cause: fire

In the case of an emergency there were two ways that a jet could provide emergency oxygen through masks to its passengers: it could carry oxygen under pressure in cylinders or, as in the case of these particular planes, it could carry oxygen-generating canisters that provide O_2 by means of a chemical reaction. The stainless-steel canisters were shaped like sausages with a pin

Flames rise from the remains of an ageing Pakistan International Airlines Fokker 27 turbo-prop that crashed 10 July 2006. Bound for the eastern city of Lahore, the aircraft plummeted to the ground minutes after take-off from Multan in central Pakistan, killing 45.

at one end. When you pulled the pin, a detonating cap would ignite sodium chlorate in the tube. The reaction releases breathable oxygen for up to twenty minutes, but the canister cases would become very hot—up to 260 degrees Celsius (500 degrees Fahrenheit). These canisters, like so many other things, had a use-by date, and mechanics were supposed to remove the old ones and fit them with safety caps so that they wouldn't accidentally trigger. Outdated, safety-capped canisters frequently found themselves as cargo on scheduled flights even though the manufacturer's manual said that the only truly safe canister was a spent canister.

Ten years before this crash a mechanic had been rummaging for spare parts in the cargo hold of a DC-10 when he accidentally triggered a canister. The fire that followed wrote off the multi-million dollar plane. The FAA knew about this problem, but didn't do anything about it.

In the days leading up to the crash mechanics at SabreTech had replaced outdated canisters from two MD-80s and replaced them with new ones. They didn't safety cap them, nor did they set them off to render them harmless. Instead they left them to clutter up the SabreTech store room. When a store clerk needed to make room, he decided to ship the canisters back to Atlanta. He wrapped them in bubble wrap with no warning labels. Handlers then placed these uncapped, unspent canisters into the front cargo hold of ValuJet 592 among some tyres.

The NTSB surmised that in the normal jostle of cargo loading or at take-off at least one retaining pin

came loose and triggered the activation of at least one canister. Since the canister wasn't attached to the emergency oxygen system the canister began to leak oxygen directly into the hold. The heat of the canister itself then ignited the bubble wrap and started a fire that may have triggered the activation of more canisters. The fire was now self-perpetuating and growing worse all the time because of the constant supply of oxygen. In the absence of a smoke detector or a fire alarm, the flight crew had no idea what was going on until the fire, seven minutes old and completely out of control, caused one of the tyres to explode as it burned through the plane's flight systems.

In the cockpit Captain Candalyn Kubeck and First Officer Richard Hazen heard a thump. The DC-9 was undergoing a total systems failure. At an altitude of 9500 feet (2890 metres), it dropped to the left and fell 8600 feet (2600 metres) in three minutes. Candalyn managed to regain control of the plane, level out at 900 feet (270 metres) and head back to Miami. Thirty-four seconds later Flight 592 tipped on its right side, turned its nose down and crashed into the Florida Everglades killing all onboard instantly. The criminal investigation found that SabreTech was guilty of transporting dangerous goods and failing to train its employees adequately.

2 September 1998

Flight: Swissair Flight 111
Model: McDonald Douglas MD-11
Registration: HB-IWF
Fatalities: 229/229 (215 passengers and 14 crew)
Principle cause: fire

Flight 111 had taken off from JFK, New York, at 20:18 bound for Geneva, Switzerland. Forty minutes into the flight, the crew smelled smoke and determined that it was confined to the cockpit. They assumed that there was a fire in the air conditioning system.

At 21:14:18, when the smoke became worse Captain Ernst Zimmerman declared, 'Pan, Pan, Pan', an international distress warning that is shorthand for: 'We have an emergency situation here, but for the moment there's no immediate danger to any person or to the vessel itself.' It's derived from the French word *panne,* meaning a mechanical breakdown. The much more famous 'Mayday'—derived from a corruption of the French *m'aidez* meaning 'help me'—is more urgent.

'Mayday' is shorthand for 'Help! We're in danger of dying or losing the ship.' When emergency services hear 'Mayday' they drop everything and come running.

In retrospect Zimmerman didn't realise what was actually happening. How could he? Nothing like what was about to happen to Flight 111 had ever happened before to any aircraft.

Zimmerman requested a diversion to Boston, an airport that he was familiar with, but which was still 1700 kilometres (1000 miles) and almost two hours away. ATC redirected him to Halifax, 370 kilometres (230 miles) and a more modest twenty minutes away. One minute after the 'Pan, Pan, Pan', the smoke was so bad that the crew had to put on oxygen masks. Ernst then began a by-the-book fire checklist scenario that would take twenty minutes in conformance with Swissair policy. What the flight crew didn't know was that they had already run out of time.

Out of the 250 kilometres (155 miles) of wiring on the plane, a single faulty wire a few inches long in the MD-11's state-of-the-art in-flight entertainment system had begun to arc. There was no cooling system installed for this arrangement, no specific fire alarm and not even a simple on-off switch. Even more critically, the wiring was in close proximity to insulation made of metallised polyethylene tetrahalate—a material that proved to be highly flammable. The fire spread so rapidly that over the next few minutes it had begun to affect display systems,

The recovered fire detection unit onboard Swissair Flight 111. Over two million pieces of the plane were recovered.

even further compromising the flight crew's ability to respond to the crisis.

By 21:19 Flight 111 was at 21 000 feet (6400 metres) and still 170 kilometres (104 miles) from Halifax. It needed time to dump fuel and lose altitude, so began a wide circle around St Margaret's Bay.

At 21:22 the cockpit was now full of smoke, even though there was no sign of a fire in the main cabin. Over the next 90 seconds, the fire reached a critical mass, breached vital lines and connections and the plane underwent a series of exponential systems failures. The autopilot disconnected and by 21:25 the cockpit was on fire. ATC in Moncton missed Ernst Zimmerman's transmission of 'we have to land immediately.'

Seconds after the last transmission, the cockpit was on fire and both the flight data recorder (FDR) and the cockpit voice recorder (CVR) failed at 10 000 feet (3000 metres). Flight 111 remained in the air and totally out of control for the next six minutes during its long fall.

The MD-11 crashed in St Margaret's Bay, Nova Scotia at 21:31. It hit the water at a speed of 555 kilometres per hour (345 miles per hour). The plane disintegrated completely.

The Transportation Safety Board of Canada then began what was to become the longest and most expensive investigation in its history. Divers, remote devices and dredges recovered over 2 million pieces of the plane from a depth of 55 metres (180 feet). The debris filled a military hangar—among it were several pieces of art, including a work by Pablo Picasso.

Representatives from the Royal Canadian Mounted Police, The National Transportation and Safety Bureau of the United States, Boeing (which by now owned McDonnell Douglas) and Swissair all joined in the investigation.

It is a measure of how important the FDR and the CVR are that authorities had to use every forensic trick in the book to reconstruct the final six minutes of Flight 111 entirely from the wreckage. It took investigators four-and-a-half years and $40 million ($50 million in today's terms) to make a final report. That's $500,000 per word to say, in essence, 'that flammable materials do not belong on aircraft'. There was nothing the crew could have done. The fire checklist book that Ernst Zimmerman referred to during the last few minutes of his life was found molten.

Swissair promptly removed all suspect materials and the metallised polyethylene was banned from aircraft but still had to endure a protracted lawsuit, as did DuPont, the manufacturer of the insulation. However the investigation went on for so long that by the time it had finished Swissair had gone bankrupt.

WORDS OF A WITNESS

ATC controller Bill Pickeral later recalled his sense of helplessness about the fate of Swissair 111:

It's a strange experience. I'm not sure that I can adequately express the feeling … You work to provide a service and you read about aircraft flying into a mountain or ending up in a swamp in some distant country by you never expect that it's going to happen in your own backyard and when it does it's kind of a lonely experience in one sense.

Despite its bursting into flames, all 297 passengers and 12 crew aboard this Air France A340 survived after it overshot the runway in Toronto, Canada on 2 August 2005.

FAMOUS CASUALTIES

AIDS researcher Jonathan Mann (52)
Chef Joseph La Motta (49), the son of boxing champion
Jake La Motta

22 August 2006

Flight: Pulkovo Aviation Enterprise Flight 612
Model: Tupolev TU-154M
Registration: RA 85185
Fatalities: 170/170 (160 passengers and 10 crew)
Principle cause: fire

Flight 612 was travelling from the Black Sea resort of Anapa to St Petersburg and was over the Russian Ukranian border when it flew into a storm cell. The storm was unusually high, climbing to 49 000 feet (15 000 metres)—well above the plane's service ceiling of 42 650 feet (13 000 metres).

The plane found itself in severe turbulence, at one point climbing from 39 000 feet (11 961 metres) to 42 000 feet (12 794 metres) in just ten seconds. At 15:37 the Tupolev sent a distress call to Moscow and reported an onboard fire. Within two minutes the plane entered a steep stall from which the crew could not recover. The plane crashed nose-first into a field near the town of Sukha Balka 45 kilometres (28 miles) north-west of Donetsk in eastern Ukraine.

Rescue workers recover bodies from a Russian Tupolev Tu-154 that crashed en route from the Black Sea resort of Anapa to St Petersburg.

The plane came down in Sukha Balka, 45 kilometres (28 miles) from the eastern Ukrainian city of Donetsk, 22 August 2006.

THE FUEL QUESTION

Why, after over a century of flying, is fire still such a killer?

In the early days of aviation, aeroplanes used the same fuel as automobile engines. These fuels had an octane rating no higher than 87. The higher the octane the more efficient the fuel but, with the invention of rotary piston engines, developers needed a fuel with more kick. Major Jimmie Doolittle, of the Aviation Fuels Section of Shell Oil Company, lobbied for the development of a fuel with an octane rating of 100. The resulting Avgas became the aircraft fuel of choice during World War II. The major problem with Avgas is its low flashpoint—at a temperature of only –1°C (30°F) it produces a vapour that ignites easily.

The development of jet engines required fuels with the following characteristics:

- Fuels with a higher flashpoint
- Fuels that didn't need to be converted into a vapour
- Fuels that didn't produce dirty, visible smoke
- Fuels less likely to create contrails— condensed water vapour or ice crystals
- Fuels that did not ignite at low temperatures

Jet Propulsion Fuel 1—JP1, was the first such fuel, but soon gave way new versions.

Jet A-1 is kerosene-based with a flashpoint above 38°C (100°F) and a freeze-point maximum (FPM) of –47°C (–52.6°F)—the main fuel used in all commercial jets, internationally. The former Eastern bloc, with its Tupolevs, use a Russian fuel—TS-1. Jet A is similar to Jet A-1, but its FPM is –40°C (–40°F) and it's only available in America. Jet B is more flammable and only used in extremely cold climates.

As wonderful as these fuels are, all have shortcomings, as so often demonstrated.

How do you create a fuel that burns fiercely when everything is going well, but becomes as non-flammable as possible when things go horribly wrong?

In the 1990s scientists developed a clever additive to solve the problem. If the fuel tanks were hit the shockwave would alter the molecules and turn the jet fuel into a low-flaming gel. It worked really well in the test tube. The American government funded an $11 million dollar experiment involving a 727 loaded with cameras, monitoring equipment and 64 crash-test dummies. Researchers deliberately crashed the remote-controlled plane onto a runway and into a wall at Edwards Air Force Base. The plane exploded into a fireball. It was a spectacular failure, but the presence of the interior cameras gave a unique insight into what it must really be like to be in an aeroplane when it's involved in a crash of this type—horrible—as you might imagine.

As usual, the military get the good stuff. The US Navy has been using low-flashpoint JP-5 since 1952. Its flashpoint is above 60°C (140°F)—20°C higher than Jet A.

The airline industry doesn't use JP-5 because it's more expensive. They don't believe the public would shell out a couple of extra bucks per trip for that added level of security. It's amazing that they don't let us decide for ourselves. With world consumption of jet fuel at 760 million litres (200 million gallons) per day, economies of scale should lower the price significantly.

In the meantime, we continue to fly with what an industry insider calls 'tombstone technology.'

ALTHOUGH THE FOLLOWING INFORMATION comes from Australian sources, the broad principles of airline safety regulation and investigation apply, theoretically at least, in an international context. Although it would be more truthful to say that these attitudes and practices are more a reflection of how people do things in the developed, industrialised, Western nations.

This at least, is not theory. The international community of aircraft regulation and safety investigation is, by necessity, a close-knit one. Planes go everywhere, they are all subject to similar manufacturing processes when they are built, they all have similar maintenance needs to keep them in the air, and the physical and psychological realities in which they operate know no national boundaries. Metal fatigue or faulty design doesn't care who you are or where you come from. A storm

PART NINE:
Air crash investigations

is a storm and a badly trained, unprepared or disoriented pilot is a badly trained, unprepared or disoriented pilot. It's therefore in the best interests of the whole aviation industry to share as much knowledge as they can about why accidents happen and how to prevent them happening again.

Apart from the occasional politicking and pettiness that seems to plague the whole human race, the general level of support, teamwork and disclosure in the regulation and accident investigation industry would be the envy of a lot of other fields of human endeavour.

You can find a great deal of comfort from knowing that, as a whole, the regulatory authorities and investigating agencies are dedicated to ensuring that air travel continues to be one of the safest ways to travel from 'A' to 'B'.

The rules of the game

The first important distinction to make is that—at least in Australia, Canada, New Zealand, South Africa, the United Kingdom and America—the people who make and enforce the rules, the regulators, are entirely separate from the people who investigate the accidents themselves.

Organisations like the US Federal Aviation Authority, the UK Civil Aviation Authority, Transport Canada, and the Civil Aviation and Safety Authority (CASA) in Australia make the rules; they set the standards and they ensure compliance with the standards. They have the legal power to enforce a change or a modification to the training of pilots, the design of aircraft or to maintenance procedures. Although the maintenance and inspection of aircraft is the responsibility of the airline, the regulators carry out spot inspections and audits. The authorities typically also carry out industry education and training and issue licences to pilots, air-worthiness certificates to individual aircraft and operator certificates to airlines. Think of them like the motor registry, only with wings.

In addition, many countries require foreign operators to hold a Foreign Operators Certificate, and it's not unusual for regulators in the developed world to ban some carriers outright from operation in their territory. (For an up-to-the-moment list of all airlines that are so bad that they're not permitted anywhere near the European Community go to: *ec.europa.eu/transport/air-ban/list_en.htm*)

Who pays for all this? CASA's budget is around AUD $130 million per year. About 17 per cent of cost recovery comes from licence fees etc. The bulk of the money comes from tax on aviation fuel.

Civil aviation is highly regulated. It has to be. Too many things can go wrong.

When things go wrong

Organisations such as the Australian Transport and Safety Board (ATSB) and the National Transport and Safety Bureau (NTSB) in America are independent government agencies. They have to be. When things go wrong, investigators need to be able to poke their noses everywhere and ask questions. To ensure that they receive the best answers, the data they collect can be used only in Coronial Enquiries. Investigators cannot be called as witnesses in criminal proceedings. When an accident does happen there are usually parallel investigations by the police or other authorities to determine if a crime has been committed. It usually becomes obvious fairly quickly if some crime was responsible for bringing down a plane. At that point it no longer becomes a matter for the safety boards, although police may call upon them to provide technical expertise.

The general mindset of safety investigators is that prior to an accident everyone was working to make things safe and everything failed. They don't play the blame game. So two things drive an investigation: an ethic of discovering the facts and rigorous scientific procedures for gathering. The methodology has evolved over decades, and since 1947 the International Civil Aviation Organisation has overseen the collaboration of 189 nations to investigate, modify, improve and enhance the process of determining the causes of aviation accidents.

Safety investigators don't measure the severity of an accident by a body count. They know that the difference between an event ending up as a report or ending up as a headline can be a very fine line—once you lose control of an aircraft you lose the guarantee of safety and survivability is often a matter of luck. What happens when the luck runs out? What happens at the site of a major hull-loss accident?

Every untoward incident involving an aeroplane requires meticulous investigation. This is how the industry improves and learns to avoid repeating its failures.

RA-85185

Ground zero: where the investigation begins, even while search and rescue teams and their dogs look for any human remains.

Ground zero

In late 2006 I spoke to representatives from the ATSB to give me their insights. The following interview was conducted with great thanks to Deputy Director Information and Investigations Alan Stray, Senior Transport Safety Investigators Robert Kells and Michael Watson, and Peter Gibson of the Australian Civil Aviation and Safety Authority.

Alan Stray: When I'm at the sight of an aircraft accident, I'm there to do a job that hopefully will help prevent future similar accidents.

Michael Watson: There's a focus that descends over my thinking processes—a rational focus that pushes other things out as I concentrate on the things that matter to me.

I first want to know who is the senior policeman there. The police, fire fighters and ambulances are usually the first to arrive. We want to know from the senior policeman as much as possible about everything that happened before we arrived ... These people want us to move forward with them to identify just what are the major sources of danger ... We don't want there to be any more victims. So we want to identify and neutralise any hazards, so that you can start bringing more people into the area.

Xavier: So step one is safety?

Alan: Safety is paramount. You might get to a site where the hull is very much intact. It's hit the ground at a relatively moderate speed, too high for survivability perhaps, but still leaving most of the wreckage in a small area. Or you might arrive at a scene that's 500 or 600 metres long. That would indicate that the aircraft has hit the ground at a high speed and at a relatively shallow angle, but you might get a lot of fragmentation with no pieces much bigger than a few square feet and many pieces much smaller in size, although more robust items like engines may be more intact, but they'd be

the biggest of the items. You could imagine then that passengers and body parts would be similarly oriented. It can be like a war zone—a huge area of mass destruction.

'It can be like a war zone—a huge area of mass destruction.'

Xavier: You've seen these sights before, but many of the people that you work with have never seen anything like this. The smell doesn't help much, does it? What's it like for the rescuers them?

Alan: It's a unique smell, difficult to describe.

Michael Watson: It's very distinctive, but often different, depending on what's been involved. Often, it's a sort of acrid plastic and oils.

Alan: A combination of fuel, oils, resins and burning, but it's mostly the plane. These days removal of the victims by the Police Disaster Victim Identification team takes place very quickly, so there's little opportunity for decomposition. Investigators and seasoned police and emergency services personnel at times find these scenes difficult. It's even harder on the less experienced officers and a little overwhelming for them, but it never ceases to amaze me how the training for these people equips them so well for what they have to do.

They look to us for guidance really early with respect to the aircraft and wreckage aspects. Part of what we do is to take into consideration what the rescuers are doing and to some extent slow things down when necessary, so no one gets hurt. We take the time to talk with them. Some people react to an accident by getting maybe a little too keen and that's when we have 'curb their enthusiasm', because once the survivors have been rescued, there's not a lot you can do for the deceased so we need to take a step back and ensure that the procedures used on site will not endanger the emergency services people and the investigators.

Xavier: You'd have to give them a gentle word about not getting too near a fuel tank that could explode any second now.

Michael: At times it would be much more than a gentle word, I can tell you! But there are other hazards that aren't in the general experience of the non-specialist.

Xavier: What are the key dangerous bits in a crashed aircraft?

Alan: You'd be looking for dangerous goods markings, for any hazards that the plane was carrying in its cargo hold. If it's a major disruption, the dangerous goods may come into contact with other dangerous goods to make a dangerous cocktail. And there are things like the oleos in landing gear containing a mixture of nitrogen gas and hydraulic oil under very high pressure. When the landing gear get damaged or burnt, these oleos could go off like a missile.

Michael: And for example the pressure of a plane tyre is much, much greater than that of a car tyre. So it's dangerous to transfer your values to what's happening with a plane.

Alan: But the new modern problem is the range of composite materials and carbon fibres that manufacturers are using to construct aircraft ... It's that cocktail of composite materials, plastics, paints, chemicals, fuels burning and generating huge billows of smoke where people are working ... The resins are carcinogenic and if the dust carried in the smoke gets into your lungs, it's the next best thing to asbestosis.

Burning isn't the only problem. When these composites fracture they form splinters as fine as acupuncture needles. If you get a needlestick injury from a composite splinter you can't pull it out, because it almost certainly will break. You have to get it excised or a splinter could get into your bloodstream with lethal results.

With all of these potential dangers, we immediately put a protection order on the accident site, and no one

can enter that site except those authorised to do so. The only exemption is for the retrieval of survivors, bodies and baggage. We may commence by noting 'witness marks'—traces of evidence—and ask people not to step there. It may take several hours before we can really get near the wreckage, but one of the main priorities is to retrieve the data recorders, because their analysis can really help to guide us to know what we're looking for. It is very much a top priority.

Xavier: What happens after you secure the safety of the area?

Alan: Well, we assign an on-site investigator-in-charge (IIC) from a roster of people on call for that work ... Other roles include: ... bureau spokesperson (usually the IIC) to deal with the media and other agencies, a family liaison officer, site safety and security, technical support.

Then you'd also assign group leaders as required:

- Fixed-wing operations and aircraft performance: to work out what the pilots did in flight and how the plane reacted to their control inputs and the expected performance.
- Site survey: to record and map out the site and note what ended up where and what that tells us about the final seconds.
- Power plants specialists: to examine the aircraft's engines.
- Systems specialists: to attend to fuel systems, electrical systems, hydraulic system, air-conditioning and pressurisation systems.
- Maintenance records examiners
- Flight data recorders and cockpit voice recorders
- Human performance specialists
- Survival factors analysts
- Cabin safety analysts

The 'shatter zone' portion of the reconstructed fuselage of Pan Am Flight 103, which exploded over Lockerbie in 1988 is kept in Edinburgh, Scotland.

Michael: As you start to investigate you learn things, and IICs can decide that they need more attention in specific areas and will assign qualified people to investigate further.

Alan: In fact the investigation team leaders can co-opt anyone they need to look more deeply into something. In a major investigation you can, and do, get a cast of hundreds involved.

The state [country] of the aircraft manufacturer will send representatives including people from the company, the regulatory agency involved. The state that manufactured the engines, the state of the operator, will do the same.

Xavier: So, literally any legitimate stakeholder in the investigation can have a representative.

Alan: We all want to know why that plane came down; what happened, why it happened, and seek to have remedial action taken to prevent it happening again.

Robert Kells: Investigators come from multi-disciplined backgrounds. For example, pilots, aircraft maintenance engineers, air traffic controllers, etc. So our teams can take on these various investigation roles. At the site itself we'd want to be looking at and preserving the ground marks, impact marks and other physical evidence, because this can be the most important

International cooperation is vital in the air crash investigation business. Australian Federal Police officers at the Boeing 737-400 crash site in Indonesia in March 2007. The pilots reportedly blamed a gust of wind. Police suspected human error.

evidence that we have—aside from the black box recorders and air traffic control recordings. We now have very sophisticated laser cameras that can build up a three-dimensional computer graphics virtual representation of a crash site.

... the evidence can be very perishable because of rain or other weather. So we also want to restrict access to the site through single path entry so that rescuers don't inadvertently destroy evidence.

Michael: And to ensure safety, we need protective clothing, and cleaning facilities.

Alan: So while all of this is happening we're also setting up command posts, liaising with emergency services and interviewing witnesses. There are also teams away from the accident site itself, going to the airport or air traffic control centres involved to gather radar and voice tapes and flight plan information and documentation. A lot happens away from the accident site.

... and we hire security guards to control the movement of people in and out of various areas and we need to create the facilities for photo IDs. We don't want the media, for example, attending a family briefing. We also need to know that people entering the crash site have blood-borne pathogen training and have been inoculated for tetanus and Hep B.

Xavier: How big can an investigation get?

Robert: As big as it has to. Imagine a series of rings around a crash site. People delivering sandwiches only get to the outer ring, but specialists with just cause get further. Fewer and fewer people can get to the centre ...

Michael: But there are also cultural imperatives. In some countries it's very important to allow families in. It's not our job to provide counselling support, but we can't be blind to their needs and concerns.

Alan: The type of support provided to the victims' families by the investigation organisation depends on the country that you're in, but in Australia it's the responsibility of the airline involved to provide family assistance, counselling, even funding to get the families there. The ATSB will provide an investigator to help, by keeping the families informed of the progress of the investigation. One of the most important things families want to know is how soon they'll get a report. We aim to get a preliminary factual report published within 30 days. Our policy is to get an interim factual report published six months after the preliminary factual report, and every six months thereafter until we release a final report.

On 7 May 2005 a Fairchild Turboprop crashed 12 kilometres north-west of Lockhart River in Queensland. The two crew and thirteen passengers were killed.

It was the greatest loss of life in an aircraft accident on Australian soil since 1968. That investigation was expected to be finalised and a report published around the end of the first quarter of 2007. The investigation process cannot be rushed.

Robert: I've always been very proud of the care that my colleagues take and the empathy they have for the families of victims. We cannot speculate as to the causes of an air crash. We try to make the process as quick and transparent as we can, but we are always at pains to check and recheck to ensure that as far as possible we don't get it wrong.

Michael: On-site sometimes is not a pleasant working environment, but it is so much more important to deal with the survivors, having to talk to them. These are people whom you can help, and who can help you, because they have their story to tell.

Xavier: Whom do you interview first? How soon do you consider it appropriate to talk to survivors?

Alan: We need to be very sensitive, particularly if they need medical treatment or psychological support. We wait. The medical professionals guide us as to when the survivors are ready to talk to us. We wait for a time that's convenient for them, in a private place. We permit them to have a person at the interview to provide comfort, that's very important. If it's a surviving pilot they may have a 'pilot's friend', who can be anyone the pilot trusts and is there to help. We tell them about the protection mechanisms in place so that they understand that their statements are restricted information for the accident investigation purpose. If there's too much distress during the interview, they can call for an adjournment.

Robert: When we interview people it's nothing like what the police do. We just let them tell us their story—what they've seen, what they know. It's totally non-threatening.

Michael: It takes me at least fifteen minutes to explain that. I want them to feel they genuinely trust me, but a lot of people we interview are witnesses who saw things that they weren't expecting to see. It's hard to get people to remember things that they weren't trying to remember at the time. The usual reaction is: 'What the **** was that?' Then there's the problem of people wondering whether or not they did anything wrong. Even if the engine was fine, the last engineer who signed off on an engine will be wondering whether she or he did something. 'Did I tighten that last screw?'

Xavier: How good is human memory?

Robert: We can't even pretend that we know all there is to know about the limitations of memory recall, but it's best to interview people as soon as possible after the accident, especially before they've had a chance to confer and come up with a common story that may not be what they really saw or heard.

Michael: People have come to me worried that they've, 'told me the wrong story', because what they've said doesn't square with what they've checked with other people.

Robert: But we always check what people say against the physical evidence, because the physical evidence is usually more reliable.

Xavier: What then is the value taking witness statements?

Robert: Because you often get consistencies that are very helpful.

Michael: When you get stories and evidence that are as separate as possible, and from different sources and different angles, and they are all telling you the same thing—where they converge, that's significant.

Xavier: What happens to the aircraft parts after you've finished with them?

Robert: They're given back to the owners, but there are other agencies like the Coroner that may need them, and there are insurance considerations too.

Alan: We are constantly working with other states [countries] to provide and exchange information. In particular, though, if we do discover something and we make a recommendation for immediate remedial action, then the organisations that the recommendations are addressed to, including foreign authorities, are asked to inform us of any action that they take.

Xavier: So people don't have to wait for the final report to get the benefit of the findings.

Michael: There is a constant flow of information to the public; interim factual reports and recommendations during the course of an investigation. I should add that it's becoming increasingly important nowadays to look at the interactions between the various organisations involved in the events that led up to the accident ...

As an industry, we're getting better at the structures of the aircraft and now we have to get better at being the people who make those structures work.

Robert: And one more important point. We often get asked, 'Who checks the checker?' Well firstly, there's a very extensive system within our organisation of peer reviewing, specialists, up and down the chain of command. Then, when everyone has agreed on what happened, we can take our findings to the coroner and they reserve the right to check the facts even further and test our findings. The investigation report is examined and analysed to the nth degree.

Alan: The draft report also goes to the directly involved parties. They have an opportunity to review our findings and they can submit any of their own findings with supporting evidence. We'll analyse and review any new evidence and change the reports if the evidence stands up to scrutiny.

All this assiduous and stringent dedication to an accurate truth as well as prompt and appropriate response helps to explain why Australia's safety record is so impressive. Certainly, the total number of passengers per year (35 482 000 for 2004–2005) is relatively small by world standards, and Australia has relatively benign weather, but there is also very strong culture of diligent vigilance. If there was ever a country that needed safe air travel it is Australia—despite the improvements in air transport, it is still a long way away from anywhere else.

Why planes crash

What have decades of investigation taught us?

It makes sense that—although flying isn't exactly unnatural—insects and birds have been doing it for millions of years—it is nevertheless decidedly tricky. So tricky that human ingenuity only successfully accomplished and barely managed to control heavier-than-air flight just over a century ago.

It isn't enough just to take to the air. Getting into the air isn't the problem; staying there and going where you want to go is the big challenge, and landing safely is the greatest challenge of all.

It doesn't help either that the medium through which planes travel isn't exactly smooth. Atmosphere has an annoying habit of creating weather. This usually involves wind, water and temperatures that turn water and wind into fog, rain, snow and ice.

Giant jigsaw puzzle. Investigators meticulously piece together the remains of the *Columbia* space shuttle.

The training of pilots, and the complex machines that help them do their job, are all designed to overcome these obstacles, but of course things still go wrong.

After over a hundred years of flying, air safety professionals have determined that the main reasons that aeroplanes crash are: aircrew error, weather and equipment failure. More specific are the reasons listed below.

(The percentages refer to the proportion of accidents involving that reason, taken from the International Civil Aviation Authorities Annual Safety Report statistics for 1995. The percentages add up to more than 100, because there is usually more than one reason for a crash.)

- *Aircrew error:* pilots, co-pilots and flight engineers misinterpret conditions, data, instructions or make a bad call. 65%.

- *Controlled flight into terrain (CFIT):* a mechanically perfect aeroplane crashes into the ground or other obstacles, because the aircrew don't see them or the instruments don't detect them and alert the aircrew in time. 37%.

- *Weather:* atmospheric conditions inimical to flying. 30%.

- *Loss of control:* the aircrew lose their ability to make the aeroplane do what they want it to do. 16%.

- *Engine fire or failure:* 14%.

Journalists in Taiwan are given the opportunity to film reconstructed wreckage of China Airlines flight 611. Metal fatigue.was suspected as the cause of the crash that killed all 225 people onboard in 2002.

- *Structural or systems failure:* this includes things like metal fatigue, fraying of wires or the degradation of materials. 11%.
- *Operations error:* doing the wrong thing at the wrong time, making an accurate assessment, but sitill pushing the wrong button or not going through the correct checking procedure. 11%.
- *Maintenance:* faulty upkeep at the ground level. Problems undetected that have later, catastrophic results. 1.7%
- *Airframe/systems fire:* 1.7%.

The trend over time is telling. According to the people at planecrashinfo.com, who have gathered statistics on 2147 civil aviation accidents from 1950 to 2004, aircrew error declined in the 1970s and 1980s and increased again in the 1990s. Weather as a major cause has declined, mainly due to improvements in how aeroplanes handle windshear and icing on the wings, but sabotage accounted for 8 per cent of crashes during the 1990s. Unfortunately, this may well increase in the future.

Equally interesting, according to *Boeing's Statistical Summary of Commercial Jet Aeroplane Accidents*, take-off, initial climb and final approach account for only 5 per cent of a plane's journey time, but account for 33 per cent of all fatalities. If you're going to die, you're much more likely to die on take-off and final approach. However, landings account for 45 per cent of all accidents, so if there's going to be a non-fatal accident, almost half of them will occur on landing.

The regions with the greatest fatalities are, in order of the most risky:

- Africa
- South America and the Caribbean
- South East Asia

'If you want the statistically safest airline to fly in, then fly Qantas. During its jet-age history of over fifty years, the Australian national carrier has never had a major hull loss.'

Emergency shutes deployed from a Qantas 747 at Sydney Airport in 2003, after a ground engineer spotted smoke from one of the jet's 16 brakes. All 347 passengers were evacuated.

- China
- India
- Eastern Europe
- The Commonwealth of Independent States (the safety record of this region is worse than when these countries were part of the USSR)
- Western Europe
- North America
- The Middle East
- Australasia

The order reflects history, geography and economics. One significant point is that the poorer regions of the world tend to get 'hand-me-down' aircraft from the richer nations. These planes are inevitably older and by definition, less safe. If you combine this with safety standards that are more lax and pilots that may not be as well trained, you have a greater likelihood of a disaster.

However, before travellers in more developed nations become too smug, they should be aware that during peak seasons airlines often rent back their older aircraft. They then repaint them temporarily with company colours, so that it's impossible for a passenger to know whether they're flying in one of the airline's usual fleet, or in a rental. This practice is known as 'wet-leasing.' Whatever the origin of the term, it sounds like a reference to the condition of the paint on the leased plane. Wet-lease contracts can last anytime from one month to two years.

If you want the statistically safest airline to fly in, then fly Qantas. During its jet-age history of over fifty years, the Australian national carrier has never had a major hull loss. Other airlines like Air Berlin are also great, but they haven't been around nearly as long.

Ultimately, though planes will continue to crash and people will continue to die, and, although fewer than one in a million passengers in the West will not reach their destination, that statistic is rather cold comfort if you or someone you love is one of those one in a million.

Airlines go to great lengths to create ever larger, more comfortable even luxurious aircraft and some idiosyncratic people even enjoy flying. For the rest of us, air travel is for the mostly a dull, somewhat uncomfortable experience and there never seems to be anything good on the in-flight radio channels. Before the *Challenger* disaster, people were even becoming blasé about space travel, and, let's face it, when's the last time you even thought about the International Space Station and what the scientists are doing up there? One day perhaps we may even reach a time when space travel too is equally dull, mundane and relatively safe. Maybe that explains our enduring fascination with air and space disasters. They shock us out of our complacency and remind us, if only for a few hours, that no matter how mundane our lives seem to be, they are nonetheless fragile and precious things and we should appreciate them, however long they last.

APPENDICES

Appendix one:
The top twelve airship disasters

According to The Airship Heritage Trust, airships have an undeserved reputation for being super dangerous. In truth the record of airships wasn't particularly bad, considering the huge amount of work that they did, the vast distances that they covered, the total number of passenger kilometres they flew and the level of material science and technology at the time.

In an effort to salvage the reputation of the airship, the table below lets the facts speak for themselves. It looks extremely tame compared to the statistics on fixed-wing, heavier-than-air craft.

Note also that the majority of fatalities are military. However, this list includes only those accidents that weren't in combat.

Rank	Date and time	Airship and nation	Location and cause	Deaths
1	04/04/1933 12:30	Akron ZRS4 America (navy)	Off the Atlantic coast, America Violent storm	73
2	21/12/1923 02:30	Dixmunde LZ114 France	Mediterranean Lightning strike and fire	52
3	05/10/1930 02:00	HMA R101 United Kingdom	East of Beauvais, France Cell rupture and fire	48
4	24/08/1921 17:35	HMA R38 UK (military)	Humbar River Estuary, UK Stress structural failure	44
5	06/05/1937 19:25	Hindenburg LZ129 Germany	Lakehurst, New Jersey, America Cell rupture, static electricity, fire	36
6	21/02/1922	Airship Roma America (army)	Hampton Roads, Virginia, America Collision with power lines, fire	34
7	17/10/1913 10:30	LZ18 Germany	Berlin, Germany Test flight engine failure, crash	28
8	30/03/1917	SL9 Germany (navy)	Baltic Sea Lightning strike, fire	23
9	03/09/1915 15:20	L10 Germany (navy)	Mouth of River Elbe, Germany Lightning strike, fire	19
10	09/09/1913 18:30	L1 Germany (navy)	Off Heligoland, Germany Violent storm	14
11	03/09/1925 05:30	Shenandoah ZR1 America	Caldwell, Ohio, America Violent storm	14
12	05/02/1938	Osoaviahim V6 USSR (navy)	Kandalaksha, USSR Human/chart error, collision with Mount Neblo	13

Appendix two:
The top 101 air disasters

The top 101 represent a total of 21 391 fatalities in air disasters over the past 50 years.

Note that these figures are only for the worst individual cases. The yearly average in the decade 1997–2006 was over 800 fatalities per year.

The worst year for air disasters so far has been 2001, with a total of four entries in the top 101, accounting for 3348 casualties. These, of course, relate to 9/11. The second worst year was 1985, with six entries in the top 102, totalling 1588 casualties. The third worst was 1996, with seven entries accounting for 1406 deaths.

Rank	Date	Flight	Location	Deaths
1	11/09/2001	American Airlines 11 and United Airlines 175	Manhattan, New York, America	2749
2	27/03/1977	KLM 4805 and Pan Am 1736	Tenerife, Canary Islands	583
3	12/08/1985	Japan Airlines 123	Mount Osutaka, Japan	520
4	12/11/1996	Saudi Arabian Airlines 763 and Kazastan 1907	New Delhi, India	349
5	03/03/1974	Turk Hava Yollari 981	Bois d'Ermenonville, France	346
6	23/06/1985	Air India 182	Atlantic Ocean, west of Ireland	329
7	19/08/1980	Saudi Air 163	Riyadh, Saudi Arabia	301
8	03/07/1988	Iran Air 655	Strait of Ormuz, Persian Gulf	290
9	19/02/2003	Islamic Revolution's Guards Corps	Shahdad, Iran	275
10	25/05/1979	American Airlines 191	Chicago, Illinois, America	273
11	21/12/1988	Pan Am 103	Lockerbie, Scotland	270
12	01/09/1983	Korean Airlines 007	West of Sakhalin Island, USSR	269
13	12/11/2001	American Airlines 587	New York, America	265
14	26/04/1994	China Air 140	Komaki, Japan	264
15	11/07/1991	Nigeria Air 2120	Jeddah, Saudi Arabia	261
16	28/11/1979	Air New Zealand 901	Mount Erebus, Antarctica	257
17	12/12/1985	Arrow Airways MF 1285R	Gander, Newfoundland, Canada	256
18	26/09/1997	Garuda 152	Buah Nabar, Indonesia	234
19	17/07/1996	TWA 800	Off New York, America	230
20	02/09/1998	Swissair 111	Off Nova Scotia, Canada	229
21	06/08/1997	Korean Airlines 801	Agana, Guam	228

22	08/01/1996	African Air Cargo	Kinshasa, Zaire	227?
23	25/05/2002	China Airlines 611	Off Penghu, Taiwan	225
24	26/05/1991	Lauda Air 004	Ban Nong Rong, Thailand	223
25	31/10/1999	EgyptAir 990	Off Nantucket, Massachusetts, America	217
26	01/01/1978	Air India 855	Off Bandra, Maharashtra, India	213
27	16/02/1998	China Airlines 676	Taipei, Taiwan	203
28	10/07/1985	Aeroflot 7425	Uch-Kuduk, Uzbekistan, USSR	200
29	04/12/1974	Martinair 138	Maskeliya, Sri Lanka	191
30	06/02/1996	Alas Nacionales (Birgenair) 301	Puerto Plata, Dominican Republic	189
31	11/09/2001	American Airlines 77	Arlington, Virginia, America	189
32	03/08/1975	Aila Royal Jordanian Airlines	Immouzer, Morocco	188
33	15/11/1978	Loftleidir	Katunavake, Sri Lanka	183
34	09/05/1987	Polskie Linie Lotnicze 5055	Warsaw, Poland	183
35	27/11/1983	Avianca 11	Madrid (Barajas), Spain	181
36	01/12/1981	Index Adria Avioproment	Mount San Pietro, Corsica, France	180
37	11/08/1979	Aeroflot 7880 and Aeroflot	Dneprodzerzhinsk, USSR	178
38	11/10/1984	Aeroflot 3352	Omsk, Russia	178
39	22/01/1973	Aila Royal Jordanian Airlines	Kano, Nigeria	176
40	10/09/1976	Index Adria Aviopromet 330 and British Airways 476	Gaj, Hrvatska, Yugoslavia	176
41	07/06/1989	Surinam Airways 764	Paramaribo, Surinam	176?
42	13/10/1972	Aeroflot	Krasnaya Polyana, USSR	174
43	04/09/1989	Cubana	Havana, Cuba	171
44	19/09/1989	Union des Transports Aériens 772	Bilma, Niger	171
45	22/08/2006	Pulkovo Aviation Enterprise 612	Donetsk, Ukraine	170
46	30/01/2000	Kenya Airways 431	Off Abidjan, Ivory Coast	169
47	31/03/1986	Mexicana 940	Maravatio, Mexico	167
48	28/09/1992	Pakistan International Airlines 268	Bhadagon, Katmandu, Nepal	167
49	07/07/1980	Aeroflot 4225	Nar Alma-Ata, Kasakastan, USSR	166
50	30/07/1971	All Nippon Airways 58 and Japanese Air Force	Morioko, Japan	163
51	16/08/2005	West Caribbean Airlines 708	La Cucharita, Venezuela	160

52	20/12/1995	American Airlines 965	Buga, Columbia	160
53	06/06/1994	China Northwest Airlines 2303	Xi'an, China	160
54	28/11/1987	South African Airways 295	Mauritius, Indian Ocean	159
55	22/12/1992	Libya Arab Airlines 1103 and Libyan Air Force	Tripoli, Libya	157
56	14/08/1972	Interflug	Königs Wusterhausen, East Germany	156
57	26/11/1979	Pakistan International Airlines	Jeddah, Saudi Arabia	156
58	16/08/1987	Northwest Airlines 255	Romulus, Michigan, America	156
59	16/03/1969	Venezolana Internacional de Aviacion 742	Maracaibo, Venezuela	155
60	03/12/1972	Spantax	Tenerife, Canary Islands	155
61	04/04/1975	US Air Force Military	Siagon, Vietnam	155
62	19/09/1976	Turk Hava Yollari	Karatepe Mountains, Turkey	154
63	29/09/2006	Gol Airlines 1907	Sao Felix do Araguaia, Brazil	154
64	09/07/1982	Pan American World Airways 759	Kenner, Louisiana, America	153
65	04/05/2002	Executive Airline Services 4226	Kano, Nigeria	149
66	19/02/1985	Iberia Airlines 610	Mt. Oiz, Spain	148
67	03/01/2004	Flash Air 604	Off Sharm el Sheikh-Ophira, Egypt	148
68	25/04/1980	Dan Air 1008	Tenerife, Canary Islands	146
69	04/07/2001	Vladivostokavia 352	Irkutsk, Russia	145
70	08/02/1989	Independent Air Inc 1851	Santa Maria, Azores	144
71	25/09/1978	Pacific Southwest 182 and Private	San Diego, California, America	144
72	17/03/1988	Avianca 410	Cucuta, Colombia	143
73	07/11/1996	Aviation Development Corporation 086	Lagos, Nigeria	143
74	23/08/2000	Gulf Air 072	Off Manama, Bahrain	143
75	05/09/2005	Mandala Airlines 091	Medan, Indonesia	143?
76	24/11/1992	China Southern Airlines 3943	Liutang, Guangxi, China	141
77	18/12/1995	Trans Service Airlift	Kahengula, Angola	141
78	29/08/1996	Vnokovo Airlines 2801	Spitsbergen, Norway	141
79	25/12/2003	Union de Transports Aeriens de Guinee 141	Cotonou, Benin	140
80	08/06/1982	VASP 168	Sierra de Pacatuba, Brazil	137

81	02/08/1985	Delta Air Lines 191	Ft. Worth-Dallas, Texas, America	135
82	16/12/1960	United Air Lines / TWA	Staten Island/Brooklyn, New York, America	134
83	04/02/1966	All Nippon Airways 8302	Tokyo Bay, Japan	133
84	08/02/1993	Iran Air and Iranian Air Force	Tehran, Iran	133
85	28/06/1982	Aeroflot 8641	Southern Belarus, USSR	132
86	19/05/1993	SAM Colombia 501	Medellin, Colombia	132
87	08/09/1994	USAir 427	Aliquippa, Pennsylvania, America	132
88	19/11/1977	TAP 425	Near Funchal on the Island of Madeira, Portugal	131
89	19/04/2000	Air Philippines 541	Samal Island, Philippines	131
90	03/06/1962	Air France	Villeneuve-le-Roi, France	130
91	16/11/1967	Aeroflot	Near Sverdlovsk, Russia	130
92	08/11/1983	TAAG Angola Airlines	Lubango, Huila, Angola	130
93	02/10/1990	Iraqi Airways	Kuwait City, Kuwait	130
94	30/06/1956	United Airlines and TWA	Grand Canyon, Arizona, America	128
95	02/10/1990	Xiamen / China SW	Guangzhou, China	128
96	21/01/1980	Iran National Airlines 'Shiraz'	Elburz Mountains, Iran	128
97	15/04/2002	Air China 129	Pusan, South Korea	128
98	09/07/2006	Sibir (S7) Airlines 778	Irkutsk, Russia	128
99	06/03/1976	Aeroflot	Voronezh, Russia	127
100	21/10/1989	TAN Airlines 414	Tegucigalpa, Honduras	127
101	23/11/1996	Ethiopian Airlines 961	Moroni, Comoros Islands	127

Appendix three:
Accident statistics of the major carriers

Airlines listed below carried in excess of 10 million passengers in 2005. (Pan Am ceased operation in 1991, but has historical importance.) Some have not operated for very long and have had fewer opportunities to suffer catastrophe. The history of some carriers is complex—mergers, acquisitions and affiliations—affecting numbers. Incidents had to be potentially lethal rather than merely inconvenient to count. Figures are based on records from 1943 to end 2006.

Carrier	Total incidents	Total fatfatalities
Air Berlin	0	0
Air China	8	129
Air France	147	1321
Air New Zealand	5	261
AirTran Airways	4	0
Alaska Airlines	15	206
Alitalia	28	46
America West	4	0
American Airlines	82	3050
American Eagle	4	0
ANA (All Nippon Airways)	22	164
Asiana	2	68
Atlantic Southeast	5	31
Austrian Airlines (Lauda Air)	1	254
British Airways (BOAC)	69 (43)	65 (392)
Cathay Pacific	8	140
China Eastern Airlines	9	40
China Southern Airlines	5	222
Comair	3	17
Continental		
Comair / Delta Connection	2	78
Delta Air Lines	51	293
Easyjet	1	0
Emirates Airlines	2	0
Expressjet	1	0

Carrier	Total incidents	Total fatfatalities
Gol Transportes Aereos	2	154
Iberia	35	583
Japan Air Lines	29	729
JetBlue Airways	1	0
KLM Royal Dutch Airlines	33	693
Korean Air Lines	23	726
Lufthansa	25	157
Lufthansa Regional	1	4
Malaysian Air Lines	4	100
Mesa	1	0
Northwest Airlines	11	18
Pan American (Pan Am)	75	1638
QANTAS	3	0
Ryan Air	2	18
SAS	21	171
Saudi Arabian Airlines	32	635
Singapore Air Lines	5	89
Skywest Airlines	4	20
Southwest Air Lines	1	0
TAM	11	145
Thai Airways / TA Int	17	382
THY (Turk Hava Yollari)	38	840
United Air Lines	63	826
US Airways	4	23
Varig	39	425

Total incidents: 915

Total fatalities: 9144

Appendix four:
The planes: profiles, specifications and fatality statistics

The following list contains the most significant commercial passenger aircraft mentioned in this book and details and specifications of interest to the casual reader.

Key to interpretation of specification details and fatality statistics

Plane name and variants: Often manufacturers will release different models to take advantage of variations in market demand. These models will have some difference in dimensions and performance from the core model. The specifications mentioned below are typical of the core model unless stated otherwise.

- **Country of origin:** Country in which the aeroplane was principally designed and manufactured.
- **Passenger capacity:** Will depend on the variation or even on alternative seating configurations designed for the same fuselage.
- **Flight deck crew:** Typically two in older propeller planes and three in older jets (pilot, co-pilot and flight engineer). The more advanced, computer-assisted planes now typically do with only a pilot and co-pilot.
- **Prototype maiden flight:** The first test flight of this model.
- **Commercial introduction:** The time of the first commercial 'maiden voyage'.
- **Number produced:** The total number of planes of this model and its variants ever

built—if the model is still in production, then the number to date.

- **Production run:** The date the manufacturer ceased to build this particular model, otherwise 'current'.
- **Type:** propeller engines, turbofan or jet.
- **Number of engines:** There usually isn't much variation in the number of engines a plane can have within a model type.
- **Maximum cruising speed:** Because of the laws of physics a plane is designed to operate at a peak level of efficiency when flying within a particular range of speed and altitude. This is the upper limit of that range.
- **Service ceiling:** The highest altitude that the plane can reach and still function.
- **Maximum range:** The farthest that the plane can go on full tanks while carrying its maximum weight in people, cargo or both.
- **Empty weight:** The weight of the bare plane without fittings, passengers or cargo.
- **Maximum take-off weight:** This is the heaviest that this particular model could ever be and still manage to leave the ground.
- **Span:** The distance from wingtip to wingtip.
- **Length:** The distance from the tip of the nose to the end of the tail, typically the tip of the rear rudder.
- **Height:** The distance from where the landing gear touches the ground to the top of the tail.
- **Wing area:** This gives you an idea of the surface area over which the air flows to create *... n* that provides the plane's lift.

Immediately below the physical characteristics of the planes are the fatality statistics. These change frequently but the listed figures are as reasonably correct as possible at the time of writing, mid 2007.

The term 'hull loss' refers to the complete destruction of or irreparable damage to an aircraft. Hull losses frequently coincide with great loss of life, but there are cases of occupied aeroplanes becoming total write-offs without anyone dying. Commercial passenger jets, for all their size and power, are rather fragile things compared to the physical forces that they may encounter.

- **Hull-loss accident deaths:** Total number of people who have died in accidents involving the hull loss of this model of aeroplane. Only a small minority of people have died in accidents without a concomitant hull loss, so this figure may be taken as a rounded sum of all fatalities in this type of aeroplane.
- **Hull losses in accidents:** Total number of hull losses involving this aeroplane type in which people have died.
- **Skyjackings and other crime occurrences:** The total number of deliberate acts of violence committed on aeroplanes of this type (including the number of hull losses resulting out of criminal events).
- **Deaths in skyjackings and other crimes:** Total deaths from deliberate acts of malice on planes of this type.
- **Non-fatal hull losses:** Total number of cases in which the aeroplane suffered total destruction or was damaged beyond repair,

but in which no one actually died.
- **Total hull losses:** Total number of aeroplanes of this type that were destroyed not only in accidents and criminal acts but also in other incidents like acts of war.

Airbus A300 / A310—B4-200—600—600R

Country of origin: France / Germany / Great Britain
Passenger capacity: 220 to 375
Flight deck crew: 2
Prototype maiden flight: 3 April 1982
Commercial introduction: 1982
Number produced: 547+
Production run: still in production
Type: turbofan
Number of engines: 2 wing-mounted
Maximum cruising speed: 965 kph (595 mph) at 10 000 m
Service ceiling: 12 000 m (39 370 ft)
Maximum range: 9175 km (5700 miles)
Empty weight: 70 275 kg (154 930 lbs)
Maximum take-off weight: 157 000 kg (346 125 lbs)
Span: 43.9 m (144 ft)
Length: 46.7 m (153 ft)
Height: 15.8 m (51.9 ft)
Wing area: 219 sq m (2 357 sq ft)
Hull-loss accident deaths: 1126
Hull losses in accidents: 17
Skyjackings and other crime occurrences: 25
Deaths in skyjackings and other crimes: 303 (3 hull losses)
Non-fatal hull losses: 3, Total hull losses: 23,
Total death toll: 1429

Airbus A320—A318—A319—A321

Country of origin: France / Germany / Great Britain

Passenger capacity: 117 to 220

Flight deck crew: 2

Prototype maiden flight: 22 February 1987

Commercial introduction: March 1988

Number produced: 2800 +

Production run: still in production

Type: turbofan

Number of engines: 2 wing-mounted

Maximum cruising speed: 965 kph (595 mph) at 10 000 m (32 800 ft)

Service ceiling: 11 887 m (39 000 ft)

Maximum range: 5300 km (3270 miles)

Empty weight: 40 150 kg (88 515 lbs)

Maximum take-off weight: 73 500 kg (161 700 lbs)

Span: 34.10 m (111 ft 10 in)

Length: up to 44.51 m (146 ft)

Height: up to 12.56 m (41 ft 2 in)

Wing area: 112.5 sq m (1210 sq ft)

Hull-loss accident deaths: 440

Hull losses in accidents: 10

Skyjackings and other crime occurrences: 5

Deaths in skyjackings and other crimes: 1

Non-fatal hull losses: 5, Total hull losses: 15

Total death toll: 441

Note: This is second most popular plane in the world after the Boeing 737.

Boeing 707

Country of origin: America

Passenger capacity: 147 to 219

Flight deck crew: 3

Prototype maiden flight: 15 July 1954

Commercial introduction: 1955

Number produced: 858

Production run: 1991

Type: turbofan

Number of engines: 4 wing-mounted

Maximum cruising speed: 974 kph (605 mph) at 9750 m (32 000 ft)

Service ceiling: 11 900 m (39 000 ft)

Maximum range: 9260 km (5750 miles)

Empty weight: 66 400 kg (146 400 lbs)

Maximum take-off weight: 151 320 kg (333 600 lbs)

Span: 44.4 m (145.75 ft)

Length: 46.6 m (153 ft)

Height: 12.9 m (42.5 ft)

Wing area: 238.35 sq m (3050 sq ft)

Hull-loss accident deaths: 2733

Hull losses in accidents: 142

Skyjackings and other crime occurrences: 60 (12 hull losses)

Deaths in skyjackings and other crimes: 289

Non-fatal hull losses: 11, Total hull losses: 165

Total death toll: 3022

Boeing 727—727-100(C)—727-200(F)

Country of origin: America

Passenger capacity: 163 to 189

Flight deck crew: 3

Prototype maiden flight: 9 February 1963

Commercial introduction: February 1964

Number produced: 1832

Production run: discontinued in January 1983

Type: turbofan

Number of engines: 3 tail-mounted

Maximum cruising speed: 920 kph (570 mph) at 7620 m (25 000 ft)

Service ceiling: 11 900 m (39 000 ft)

Maximum range: 4000 km (2485 miles)

Empty weight: 46 700 kg (103 000 lbs)

Maximum take-off weight: 95 025 kg (209 500 lbs)

Span: 32.9 m (108 ft)

Length: 46.7 m (153 ft 3 in)

Height: 10.35 m (34 ft)

Wing area: 158 sq m (1700 sq ft)

Hull-loss accident deaths: 3704

Hull losses in accidents: 87

Skyjackings and other crime occurrences: 182 (3 hull losses)

Deaths in skyjackings and other crimes: 345

Non-fatal hull losses: 11, Total hull losses: 101

Total death toll: 4049

Note: The 727 was the first trijet introduced into commercial service and it was the best-selling airliner in the first 30 years of the jet-age (1952–1982), with 1831 sold to 101 customers.

Boeing 737—300

Country of origin: America

Passenger capacity: 130 to 149

Flight deck crew: 3

Prototype maiden flight: 8 August 1967

Commercial introduction: 1968

Number produced: 1113

Production run: 1999

Type: turbofan

Number of engines: 2 wing-mounted

Maximum cruising speed: 925 kph (565 mph) at 9145 m (30 000 ft)

Service ceiling: 11 900 m (39 000 ft)

Maximum range: 4265 km (2650 miles)

Empty weight: 27 700 kg (609 400 lbs)

Maximum take-off weight: 53 070 kg (117 000 lbs)

Span: 28.35 m (93 ft)

Length: 30.5 m (100 ft)

Height: 11.3 m (37 ft)

Wing area: 91 sq m (980 sq ft)

Hull-loss accident deaths: 718

Hull losses in accidents: 14

Skyjackings and other crime occurrences: 1

Deaths in skyjackings and other crimes: 0

Non-fatal hull losses: 0, Total hull losses: 14

Boeing 747—SP—100—200—300—400

Country of origin: America

Passenger capacity: 440 to 660

Flight deck crew: 3

Prototype maiden flight: 9 February 1969

Commercial introduction: 1969

Number produced: 1369

Production run: still in production

Type: turbofan

Number of engines: 4 wing-mounted

Maximum cruising speed: 940 kph (585 mph) at 9150 m (30 000 ft)

Service ceiling: 13 715 m (45 000 ft)

Maximum range: 9625 km (5980 miles)

Empty weight: 171 000 kg (377 000 lbs)

Maximum take-off weight: 365 150 kg (805 000 lbs)

Span: 59.65 m (195 ft 8 in)

Length: 70.5 m (231 ft 4 in)

Height: 19.35 m (63 ft 6 in)

Wing area: 511 sq m (5500 sq ft)

Hull-loss accident deaths: 2850

Hull losses in accidents: 37

Skyjackings and other crime occurrences: 34

Deaths in skyjackings and other crimes: 882 (4 hull losses)

Non-fatal hull losses: 2, Total hull losses: 43

Boeing 757

Country of origin: America

Passenger capacity: 201 to 240

Flight deck crew: 2

Prototype maiden flight: February 1982

Commercial introduction: 1982

Number produced: 1050

Production run: ended 2004

Type: jet

Number of engines: 2 wing-mounted

Maximum cruising speed: 965 kph (595 mph) at 9150 m (30 000 ft)

Service ceiling: 11 900 m (39 000 ft)

Maximum range: 7315 km (4550 miles)

Empty weight: 57 040 kg (125 750 lbs)

Maximum take-off weight: 100 000 kg (220 000 lbs)

Span: 38 m (125 ft)

Length: 47.5 m (155 ft 6 in)

Height: 13.6 m (44 ft 6 in)

Wing area: 181.25 sq m (1950 sq ft)

Hull-loss accident deaths: 6

Hull losses in accidents: 467

Skyjackings and other crime occurrences: 9

Deaths in skyjackings and other crimes: 108

Non-fatal hull losses: 0, Total hull losses: 6

Boeing 767

Country of origin: America

Passenger capacity: 245 to 304

Flight deck crew: 2 or 3

Prototype maiden flight: 26 September 1981

Commercial introduction: 1982

Number produced: 934 +

Production run: still in production

Type: turbofan

Number of engines: 2 wing-mounted

Maximum cruising speed: 965 kph (595 mph) at 9150 m
(30 000 ft)

Service ceiling: 10 725 m (35 200 ft)

Maximum range: 11 390 km (7080 miles)

Empty weight: 79 380 kg (175 000 lbs)

Maximum take-off weight: 172 365 kg (380 000 lbs)

Span: 47.6 m (156 ft)

Length: 54.9 m (180 ft 3 in)

Height: 15.8 m (52 ft)

Wing area: 283.35 sq m (3050 sq ft)

Hull-loss accident deaths: 569

Hull losses in accidents: 6

Skyjackings and other crime occurrences: 7

Deaths in skyjackings and other crimes: 282 (2 hull losses)

Non-fatal hull losses: 0, Total hull losses: 11

Concorde

Country of origin: France and United Kingdom
joint venture

Passenger capacity: up to 144

Flight deck crew: 3

Prototype maiden flight: 2 March 1969

Commercial introduction: 1970

Number produced: 20

Production run: 1979

Type: turbojet

Number of engines: 4 wing-mounted

Maximum cruising speed: Mach 2.04—2180 kph
(1355 mph) at 15 635 m (51 000 ft)

Service ceiling: 18 290 m (61 000 ft)

Maximum range: 6230 km (3870 miles)

Empty weight: 78 700 kg (173 500 lbs)

Maximum take-off weight: 185 100 kg (408 000 lbs)

Span: 25.55 m (83 ft 10 in)

Length: 62.1 m (203 ft 8 in)

Height: 11.4 m (37 ft 6 in)

Wing area: 358 sq m (3855 sq ft)

Hull-loss accident deaths: 109

Hull losses in accidents: 1

Skyjackings and other crime occurrences: 0

Deaths in skyjackings and other crimes: 0

Non-fatal hull losses: 0, Total hull losses: 1

De Havilland Comet 1, 1A, 2, 3, 4, 4B, 4C

Country of origin: United Kingdom

Passenger capacity: up to 119

Flight deck crew: 3

Prototype maiden flight: 1949

Commercial introduction: 1951

Number produced: 112

Production run: 1962

Type: jet

Number of engines: 4 wing-mounted

Maximum cruising speed: 805 kph (500 mph)

Service ceiling: 11 887 m (39 000 ft)

Maximum range: 6900 km (4290 miles)

Emp   t: 36 110 kg (79 608 lbs)

Ma   ff weight: 73 480 kg (162 000 lbs)

Spa

APPENDICES

Height: 8.7 m (28 ft 6 in)

Wing area: 197 sq m (2120 sq ft)

Hull Losses: 26

Hull-loss accident deaths: 496

Hull losses in accidents: 20

Skyjackings and other crime occurrences: 5

Deaths in skyjackings and other crimes: 66

Non-fatal hull losses: 1, Total hull losses: 26

Note: The specifications above are for the last Comet model, the 4C. Earlier models were slightly smaller. The maximum take-off weight for the 1A was 52 160 kg, its range only 2414 km. In 1960 you would have paid about £1.25 million for a 4C (around $100 million in today's terms).

Ilyushin 62 Classic

Country of origin: Soviet Union

Passenger capacity: 174 up to 186

Flight deck crew: 5

Prototype maiden flight: January 1963

Commercial introduction: 1963

Number produced: 186

Production run: ended 1993

Type: turbofan

Number of engines: 4 tail-mounted

Maximum cruising speed: 900 kph (560 mph) at 10 000 m (32 800 ft)

Service ceiling: 12 800 m (42 000 ft)

Maximum range: 7800 km (4850 miles)

Empty weight: 69 400 kg (152 700 lbs)

Maximum take-off weight: 165 000 kg (363 000 lbs)

Span: 43.2 m (141 ft 8 in)

Length: 53.1 m (174 ft 4 in)

Height: 12.35 m (40 ft 6 in)

Wing area: 282.2 sq m (3040 sq ft)

Hull-loss accident deaths: 1066

Hull losses in accidents: 18

Skyjackings and other crime occurren

Deaths in skyjackings and other c

Non-fatal hull losses: 3

Lockheed Constellation (Model 1649 G - Starliner)

Country of origin: America

Passenger capacity: up to 180

Flight deck crew: 3

Prototype maiden flight: 1943

Commercial introduction: 1945

Number produced: 856 (331 military, 525 commercial)

Production run: 1967

Type: turbofan

Number of engines: 4

Maximum cruising speed: 606 kph (377 mph) at 5 669 m (18 600 ft)

Service ceiling: 7223 m (23 700 feet)

Maximum range: 7950 km (4940 miles)

Empty weight: 41 969 kg (91 645 lbs)

Maximum takeoff weight: 72 575 kg (160 000 lbs)

Span: 45.72 M (150 ft)

Length: 35.41 M (116 ft)

Height: 7.54 M (24.75 Ft)

Wing area: 171.81 Sq metres (1850 sq ft)

Hull-loss accident deaths: 85

Hull losses in accidents: 8

Skyjackings and other crimes: 1

Deaths in skyjackings other crimes: 78 (1 hull loss)

Non-Fatal Hull Losses: 0, Total Hull Losses: 10

Trivia – The Constellation came in several models. The L-1049 Super Constellation was used both for military and commercial applications. It's accident record is rather grimmer than for the purely commercial Starliner.

Hull-loss accident deaths: 1063

Hull losses in accidents: 98

Skyjackings and other crimes: 2

Deaths in skyjackings and other crimes: 31 (2 hull losses)

Non-fatal hull losses: 11, Total hull losses: 113

Lockheed Electra L-188—The Electra—Series A and C

Country of origin: America

Passenger capacity: 80 up to 99

Flight deck crew: 2 / 3

Prototype maiden flight: 6 December 1957

Commercial introduction: 1958

Number produced: 222

Production run: 170

Type: turbo-prop

Number of engines: 4 wing-mounted

Maximum cruising speed: 650 kph (405 mph) at 7620 m (25 000 ft)

Service ceiling: 8655 m (28 400 ft)

Maximum range: 3540 km (2200 miles)

Empty weight: 26 000 kg (57 200 lbs)

Maximum take-off weight: 51 250 kg (112 750 lbs)

Span: 30 m (99 ft)

Length: 32.15 m (104 ft 6 in)

Height: 10.25 m (32 ft 8 in)

Wing area: 121 sq m (1300 sq ft)

Hull-loss accident deaths: 1041

Hull losses in accidents: 47

Skyjackings and other crime occurrences: 8

Deaths in skyjackings and other crimes: 2 (1 hull loss)

Non-fatal hull losses: 9, Total hull losses: 57

Total death toll: 1043

Lockheed TriStar L-1011

Country of origin: America

Passenger capacity: Up to 400

Flight deck crew: 2 / 4

Prototype maiden flight: 17 November 1970

Commercial introduction: 1971

Number produced: 250

Production run: 1983

Type: turbofan

Number of engines: 3

Maximum cruising speed: 975 kph (605 mph) at 9150 m (30 000 ft)

Service ceiling: 12 800 m (42 000 ft)

Maximum range: 9655 km (6000 miles)

Empty weight: 109 300 kg (241 000 lbs)

Maximum take-off weight: 225 000 kg (496 000 lbs)

Span: 47.35 m (155 ft 4 in)

Length: 50 m (164 ft)

Height: 16.9 m (55 ft 4 in)

Wing area: 321 sq m (3456 sq ft)

Hull-loss accident deaths: 548

Hull losses in accidents: 7

Skyjackings and other crime occurrences: 15

Deaths in skyjackings and other crimes: 14

Non-fatal hull losses: 2, Total hull losses: 25

McDonnell Douglas DC-8

Country of origin: America

Passenger capacity: 259 up to 280

Flight deck crew: 2 / 3

Prototype maiden flight: 30 May 1958

Commercial introduction: 1959

Number produced: 556

Production run: 1972

Type: turbofan

Number of engines: 4 wing-mounted

Maximum cruising speed: 855 kph (530 mph) at 10 650 m (35 000 ft)

Service ceiling: 13 800 m (42 000 ft)

Maximum range: 7500 km (4660 miles)

Empty weight: 73 800 kg (162 700 lbs)

Maximum take-off weight: 147 415 kg (325 000 lbs)

Span: 45.3 m (148 ft 6 in)

Length: 57.1 m (187 ft 6 in)

Height: 12.9 m (42 ft 6 in)

Wing area: 272 sq m (2927 sq ft)

Hull-loss accident deaths: 2257

Hull losses in accidents: 74

Skyjackings and other crime occurrences: 48 (2 hull losses)

Deaths in skyjackings and other crimes: 75

Non-fatal hull losses: 6, Total hull losses: 82

McDonnell Douglas DC-9 / MD 80 / MD 81

Country of origin: America

Passenger capacity: 137

Flight deck crew: 2 / 3

Prototype maiden flight: 19 October 1979

Commercial introduction: 1980

Number produced: 1194

Production run: ended 1999

Type: turbofan

Number of engines: 2 tail-mounted

Maximum cruising speed: 850 kph (530 mph) at 10 650 m (35 000 ft)

Service ceiling: 13 900 m (42 000 ft)

Maximum range: 2900 km (1800 miles)

Empty weight: 35 630 kg (78 550 lbs)

Maximum take-off weight: 63 000 kg (138 600 lbs)

Span: 32.8 m (107 ft 11 in)

Length: 45 m (147 ft 11 in)

Height: 9 m (29 ft 8 in)

Wing area: 118 sq m (1270 sq ft)

Hull-loss accident deaths: 876

Hull losses in accidents: 18

Skyjackings and other crime occurrences: 6

Deaths in skyjackings and other crimes: 0

Non-fatal hull losses: 3, Total hull losses: 27

McDonnell Douglas DC-10

Country of origin: America

Passenger capacity: 380

Flight deck crew: 2 / 3

Prototype maiden flight: 29 August 1970

Commercial introduction: August 1971

Number produced: 446

Production run: 1989

Type: turbofan

Number of engines: 3 (2 wing-mounted, 1 tail-mounted)

Maximum cruising speed: 908 kph (546 mph) at 9150 m (30 000 ft)

Service ceiling: 10 180 m (33 400 ft)

Maximum range: 7410 km (4600 miles)

Empty weight: 121 200 kg (267 200 lbs)

Maximum take-off weight: 263 100 kg (580 000 lbs)

Span: 50.4 m (165 ft 4 in)

Length: 55.5 m (182 ft)

Height: 17.7 m (58 ft)

Wing area: 368 sq m (3960 sq ft)

Hull-loss accident deaths: 1432

Hull losses in accidents: 26

Skyjackings and other crime occurrences: 9

Deaths in skyjackings and other crimes: 172

Non-fatal hull losses: 2, Total hull losses: 38

Tupolev 154—Careless

Country of origin: Soviet Union

Passenger capacity: up to 180

Flight deck crew: 3

Prototype maiden flight: 1971

Commercial introduction: 1972

Number produced: 923 +

Production run: still in production

Type: turbofan

Number of engines: 3 tail-mounted

Maximum cruising speed: 900 kph (560 mph) at 12 000 m (39 370 ft)

Service ceiling: 13 000 m (42 650 ft)

Maximum range: 2750 km (1700 miles)

Empty weight: 50 775 kg (111 950 lbs)

Maximum take-off weight: 94 000 kg (207 235 lbs)

Span: 37.5 m (123 ft 2 in)

Length: 47.9 m (157 ft 1 in)

Height: 11.4 m (37 ft 6 in)

Wing area: 207.5 sq m (2169 sq ft)

Hull-loss accident deaths: 2470

Hull losses in accidents: 55

Skyjackings and other crime occurrences: 35

Deaths in skyjackings and other crimes: 245 (4 hull losses)

Non-fatal hull losses: 3, Total hull losses: 64

Appendix five:
Surviving

After reading extensively through scenario after scenario of air crashes you might be wondering what you can do to increase your chance of living through such a disaster. To be honest, there's not much, but every little bit helps and here are some basic tips:

Since most accidents occur during the take-off/climb and the descent/landing phases of a trip the fewer of these that you experience the better. If you have a choice, choose to fly on a non-stop route.

This book is about the big disasters. Despite the impression it may give, more accidents occur on smaller aircraft, so when you can, travel in big planes.

Actually listen to and understand the safety instructions at the beginning of the flight and be prepared to implement their recommended course of action in an emergency.

Hand luggage weight restrictions are there for a reason. Overhead storage bins are not the place to keep your set of free weights. In a crash heavy objects falling from overhead bins do serious damage.

Don't bring hazardous material with you. There are special provisions for transporting hazardous substances by air. Use them.

Keep your belt fastened when you're seated. You never know when you might hit an air pocket and the plane suddenly (and, hopefully, temporarily) loses altitude.

Listen to the flight attendants. They're usually better informed than you are about what to do in an emergency.

Alcohol and flying don't mix. It takes less booze to make you tipsy at altitude than at sea level. In an emergency you'll need every brain cell fully firing.

It doesn't really matter where you sit, so don't worry about it. There are scenarios in which any seat is safer than the others and there is only a slight statistical advantage in sitting towards the tail end of the plane.

Count the number of seats between you and the nearest exit. In a smoke-filled cabin, you'll be effectively blind so you'll have to feel your way out and you don't want to miss an open exit just because you can't see it.

If you land in water or if you think that you're going to land in water don't inflate your lifejacket before you're out of the plane because if the plane sinks, and you're still on it, you'll be trapped if the cabin floods.

Leave your luggage behind. It will be covered by insurance and if you have a laptop that your whole world depends on then make a back-up before you leave and store it in your office or home.

Be prepared to take responsibility for saving yourself. Don't assume that the cabin crew or flight crew are fully informed and don't let the cabin crew assume that the flight crew are fully informed either. Speak up if you see something and make sure that people listen. Take charge if you have to.

Appendix six:
References and recommended reading

There is a huge amount of information out there for air and space disaster aficionados. Feel free to consult any of the sources listed below. I recommend the following websites, they contain valuable and interesting facts and figures:

- airdisaster.com – is a nice all-rounder for accidents after 1950.
- planecrashinfo.com – is particularly useful for statistics and profiles on accidents prior to 1943.
- aviation-safety.net/index.php – the Aviation Safety Network's site has an excellent database of profiles after 1943 and many of the profiles feature voice transcripts.

Internet references

aero-defense.ihs.com/products/regulations/aviation-av-data/uk-civil-aviation-authority.htm
aviation-safety.net/index.php
dnausers.d-n-a.net/dnetGOjg/Research.htm
en.wikipedia.org/wiki/Gol_Transportes_A%C3%A9reos_Flight_1907
en.wikipedia.org/wiki/Joe_Sharkey
en.wikipedia.org/wiki/Plane_crash
en.wikipedia.org/wiki/PSA_Flight_182
en.wikipedia.org/wiki/Roald_Amundsen
info.detnews.com/history/story/index.cfm?id=33&category=government
news.bbc.co.uk/2/hi/south_asia/2811855.stm
news.bbc.co.uk/onthisday/default.stm
news.bbc.co.uk/onthisday/hi/witness/october/19/newsid_3112000/3112466.stm
spaceflightnow.com/shuttle/sts107/status.html
www.aaib.dft.gov.uk/publications/formal_re

92_502702.cfm
www.aaib.gov.uk/sites/aaib/home/index.cfm
www.aircraft-photos.net/BA.aspx
www.airliners.net/articles/read.main?id=90
www.airsafe.com/
www.airshipsonline.com/airships/r100/index.html
www.airshipsonline.com/airships/r101/Crash/R101_Crash.htm
www.arlingtoncemetery.net/
www.atsb.gov.au/
www.atwonline.com/channels/dataAirlineEconomics/index.html
www.aviationexplorer.com/
www.avitop.com/aviation/avitop.asp?Category=c
www.baaa-acro.com/
www.bmj.com/cgi/content/full/320/7237/768
www.caa.co.za/About%20Us.htm
www.casa.gov.au/
www.cbc.ca/news/background/airindia/
www.centennialofflight.gov/essay/Explorers_Record_Setters_and_Daredevils/quimby/EX5.htm
www.centennialofflight.gov/essay/Government_Role/security/POL18.htm
www.crimelibrary.com/notorious_murders/mass/jack_graham/index.html
www.crimelibrary.com/notorious_murders/mass/jack_graham/index.html
www.damninteresting.com/?p=18
www.entheology.org/edoto/anmviewer.asp?a=3&z=1
www.flight182.com/index.htm
www.geocities.com/CapeCanaveral/Lab/8803/
www.geocities.com/donuts13/index.htm
www.govt.nz/record?recordid=68
www.gutenberg.org/files/16130/16130-h/16130-h.htm

www.hq.nasa.gov/office/pao/History/sts51l.html

www.iata.org/pressroom/wats/wats_passenger_flown.htm

www.icao.int/

www.imdb.com

www.jetphotos.net/showphotos.php

www.jetpsa.com/memorial/memorial.html

www.mackenzieinstitute.com/2002/terror100602.htm

www.maurymuehleisen.com/about.html

www.mishalov.com/wtc_flight-11-transcript.html

www.nasa.gov/columbia/home/index.html

www.nationaudio.com/News/EastAfrican/10042000/
Regional/Regional14.html

www.newyorkmetro.com/nymetro/news/sept11/2003/
n_9189/index.html

www.ntsb.gov/

www.nytimes.com/2006/10/03/business/03road.html?e
x=1317528000&en=5757e7dc53790190&ei=5088&part
ner=rssnyt&emc=rss

www.panamair.org/accidents/accidents.htm

www.planecrashinfo.com/

www.rafmuseum.org.uk/milestones-of-flight/british_
civil/1920.html

www.smithsonianmagazine.com/issues/2006/august/
object.php

www.spacefacts.de/

www.tc.gc.ca/air/menu.htm

www.timetableimages.com/index.htm

www.usatoday.com/news/sept11/2002-08-12-
clearskies_x.htm

www.usdoj.gov/opa/pr/2006/March/06_crm_172.html

www.world-airport-codes.com/

www1.korea-np.co.jp/pk/073rd_issue/98120901.htm

Internet News Agencies and Services

ABC, Associated Press, BBC, CBC, CNN, Crime
Channel, Fox, MSBC, Reuters, USA Today.

Magazines and periodicals in print and online

Accident Digests

Airlines, World Transport Press, South Florida, Florida.

Airliners magazine

Airways Magazine, Airways International Inc., Sandpoint,
Idaho

Aviation History magazine

Aviation Week and Space Technology magazine

Chicago Tribune newspaper

Civil Aeronautics Board Accident Investigation reports

International Civil Aviation Organization ICAO Adrep
Summary 1988–1994

International Civil Aviation Organization ICAO Circular
Aircraft

Los Angeles Herald-Examiner newspaper

Los Angeles Times newspaper

Lost Birds magazine

Markowitz, A, 29 December 1992, 'Memory', *The Miami
Herald* newspaper

New York Times newspaper

NTSB Reporter, Monthly Digest, White Plains, NY..

National Transportation Safety Board, Accident
Investigation Reports

National Transportation Safety Board, Accident Synopses

Transportation Safety Board of Canada, Reports

Washington Post newspaper

Books

Adair, B 2004, *The Mystery of Flight 427*, Smithsonian
Institution.

Aircraft Accident Investigation Board Reports, AAIB, UK.

Barlay, S 1970, *The Search For Air Safety*, William
Morrow and Company Inc.

Barlay, S 1990, *The Final Call*, Pantheon Books.

Bartelski, J 2002, *Disasters in the Air*, Airlife.

Biggs, D 1978, Pressure Cooker, George J McLeod Ltd.

life's Register of Aircraft Accidents,

Brookes, A 1992, *Disaster in the Air*, Ian Allan Publishing..

Brookes, A 1996, *Flights to Disaster,* Ian Allan Publishing.

Byrne, G 2002, *Flight 427: Anatomy of an Air Disaster*, Copernicus Books.

Chiles, JR 2001, *Inviting Disaster*, Harper Collins, New York.

Cobb, RW and Promo, DM 2003, *The Plane Truth*, Brookings Institute Press.

Collins, RL 1992, *Air Crashes*, Thomasson-Grant.

Coote, R 1993, *Air Disasters*, Wayland Publishers Ltd.

Cushing, S and Eddy, P 1994, *Fatal Words: Communication Clashes and Aircraft Crashes*, University of Chicago Press.

Denham, T 1996, *World Directory of Airliner Crashes*, Stephens Patrick Ltd.

Donald, D (gen. ed.) 1999, *The Encyclopedia of Civil Aircraft*, Orbis Publishing Ltd/Brown Books, London.

Edwards, A 1993, *Flights into Oblivion*, Paladwar Press.

Elder, R and S 1977, *Crash*, Atheneum, New York.

Faith, N 1997, *Black Box*, The Bath Press.

Fuller, JG 1978, *The Ghost of Flight 401*, Berkeley Publishing Group.

Gero, D 1996, *Aviation Disasters: The world's major civil airliner crashes since 1950*, 2nd ed, Patrick Stephens Ltd.

Gero, D 1997, *Flights of Terror*, Patrick Stephens Ltd.

Gero, D 1999, *Military Aviation Disasters*, Patrick Stephens Ltd.

The Greatest Disasters of the 20th Century, Marshall Cavendish Ltd, London 1989.

Haine, Col EA 2000, *Disaster in the Air*, Cornwall Books.

Hui, KH 1993, *The Tears of My Soul*, William Morrow.

Job, M 1994–1999, *Air Disaster,* vols 1, 2, 3, 4, Aerospace Publications Pty Ltd.

Klee, U 2001, *JP Airliner Fleets 2001/2002*, Bucher & Co.

Knight, C & KS 2002, *Plane Crash*, Chilton Company Press.

Krause, S 1996, *Aircraft Safety, Accident, Investigations, Analyses and Applications*, Lowell, V 1968, *Airline Safety is a Myth*, Bantam..

Macha, GP 1997, *Aircraft Wrecks in the Mountains and Deserts of California*, Info Net Publishing

MacPherson, M (ed) 1998, *The Black

Collins, London.

Nader, R and Smith, W 1994, *Collision Course*, Tab Books.

Nash, JR 1976, *Darkest Hours*, Burnham.

Oster, C, Strong, J and Zorn, C 1992, *Why Planes Crash*, Oxford University Press.

Owen, D 1998, *Air Accident Investigation*, Haynes Publishing.

Parrado, N 2006, *Miracle in the Andes,* Crown Publishers.

Potter, E and Page, B 1976, *Destination Disaster*, Times Newspapers Ltd.

Reid, PP 1974, *Alive*, JP Lippincott Company.

Serling, RJ 1964, *The Probable Cause*, Doubleday.

Serling, RJ 1969, *Loud and Clear*, Dell Publishing.

Sheridan and Thomas 1994, *Survivors: Lockerbie*, PanMacmillan.

Stewart, S 1994, *Air Disasters*, Barnes and Nobel Inc.

Stich, R 1990, *Unfriendly Skies*, Diablo Western Press Inc.

Walters, JM and Sumwalt, RL III 2000, *Aircraft Accident Analysis: Final Reports*, McGraw Hill.

World Aircraft Accident Summary, Airclaims UK.

World Airline Accident Summary: 1946–1972, British Civil Aviation Authority.

World Commercial Aircraft Accidents 1946–1992, Lawrence Livermore National Laboratory, 1993.

Veronico, N, Davies, E, Mccomb, D Jr and Mccomb, M 1998, *Wreck Chasing*, World Transport Press Inc.

Video references

Air Disasters: The Facts, Visual Corporation Ltd, 1996.

Flight 401, Charles Fries Productions, 1978.

Mayday, Seasons 1,2 and 3, Cineflix Productions/ National Geographic, 2003-2006.

INDEX

Lufthansa, 70
Lynyrd Skynyrd, 215–16

M

McDonald, Lawrence, 55
maintenance error, 120, 135, 147, 174, 191, 254, 256–7, 276
Manchester United Football Team, 211
Mandala Airlines Flight 091, 101
Marciano, Rocky, 214
Martin, Dean Paul 'Dino', 217
Martinair Holland Flight 138, 113
mechanical failure, 93, 101, 139, 144, 164, 236, 253
Mexicana Flight 940, 135
midair collision, 96, 166–71, 177, 182–5, 188, 192, 196–205
Middle Eastern Airlines Flight 265, 96
military aircraft
    B-25D bomber, 35
    F-86 Sabre jet fighters, 176
    Heinkel He dive bomber, 88
    Lockheed Hercules, 64–5
    Sukhoi Su-15 fighter, 53
Miller, Glenn, 210
missile attack, 53–5, 62–5
Mohammed, Khalid Shaikh, 85
Mohawk Airlines Flight 40, 174
Morrow, Vic, 216–17
Mount Erebus disaster, 123–5
Moussaoui, Zacharias, 84
murder of flight crew, 52, 61–2
Murphy, Audie, 214

N

9/11, 82–5
Nationair Flight 2120, 275
National Airlines
    Flight 27, 250
    Flight 967, 50
    Flight 2511, 50
National Transport and Safety Bureau (USA), 286
navigation error, 94, 113, 123, 127, 131, 148
Nelson, Ricky, 217
Nielsen, Peter, 199–200, 203
Northwest Airlines Flight 305, 243
Norway (Shute), Nevil, 25
Ntaryamira, President Cyprien, 218

O

Onassis, Alexander, 214

P

Pacific Airlines Flight 773, 52
Pacific Southwest Airlines
    Flight 182, 182–3

Flight 1771, 61–2
Pakistan Air Force, 65
Pakistan International Airlines
    Flight 268, 144
    Flight 740, 272
Pan American World Airways (Pan Am), 70–1
    Flight 73, 71
    Flight 103, 65–9
    Flight 110, 70
    Flight 214, 173
    Flight 759, 108
    Flight 830, 70
    Flight 843, 236
    Flight 1736, 114–19
Parmar, Taliwinder Singh, 56–9
Parrado, Nando, 102–5
Picatinny Liquid Explosive (PLX), 60
pilot or crew error, 93–4, 96, 98, 100–1, 113, 119, 120, 123, 127–8, 131, 132, 136, 139, 144–5, 151, 159–60, 164–5, 177, 186, 189, 236, 239, 250–1, 254, 257–8, 260
pitch, definition, 7
Pitre, Marguerite, 43
Powers, Francis Gary, 215
Pulkovo Aviation Enterprise Flight 612, 282
Pruss, Captain Max, 31
Puhlmann, Rico, 152

Q

Qantas Flight 1, 260
Quimby, Harriet, 208

R

Redding, Otis, 214
Reeves, Jim, 213
Reyat, Inderjit Singh, 56–9
roll, definition, 7
Rolls, Charles Stewart, 208
Royal Airship Works, 25
Ruest, Généreux, 43
runway collision, 114–19

S

sabotage, 40–50, 56, 59–60, 65, 144
Sagal, Boris, 216
St Petersburg-Tampa Airboat Line, 21
SAM Colombia Flight 501, 144
Saudi Arabian Airlines
    Flight 163, 272–3
    Flight 763, 192–5
Scandinavian Airlines (SAS) Flight 686, 196–9
seaplane bi-plane, 24
Selfridge, Thomas Etholen, 16–18
September 11, 2001, 82–5
'shoe bomber', 261–3
Sibir (S7) Airlines Flight 778, 165

INDEX